Grim Reapers

French Escadrille 94 in World War I

Jon Guttman

Grim Reapers

French Escadrille 94 in World War I

Jon Guttman

This decades-in-the-making squadron history is dedicated to the memories of André Martenot de Cordoux, Pierre Chan, Pierre Marinovitch, Houston Woodward, Austen Ballard Crehore and all of the pilots and ground support personnel who served in *Escadrille* 94 at one time or another during the Great War.

Acknowledgments

The author wishes to thank the following—some of whom are no longer with us—for their invaluable assistance in getting this story to completion: *Escadrille* 94 member André Martenot de Cordoux, his daughter Élizabeth Martenot de Cordoux, *Escadrille* 94 member Pierre Chan, the *Service Historique de Défence (Air)*, Frank W. Bailey, Christophe Cony, Albin Denis, Richard Duiven, Jack Eder, June Gulick, daughter of Austen B. Crehore, his grandson, Jack Gulick and his granddaughter, Katherine Hoffman, Bernard Klaeylé, Philippe Lagnier, David Méchin, Mike O'Neal, Steve St. Martin, Steve Suddeby, Alan Toelle, Harry Van Dorssen and Greg VanWyngarden. This Update 1 adds a small anecdote about Austen B. Crehore.

Cover

The cover painting by Steve Anderson depicts *Sous-Lieutenant* Pierre Marinovitch's victory over a Fokker D.VII on 27 October 1918. This victory, Marinovitch's 20th, probably ended the career of German ace *Vzfw.* Karl Schlegel of *Jasta* 45 (22 victories) who was KiA on that date.

Design and layout: Jack Herris
Cover painting: Steve Anderson
Cover design: Aaron Weaver
Color Profiles: Bob Pearson
Digital photo editing: Aaron Weaver & Jack Herris

Aeronaut Books

www.aeronautbooks.com

Publisher's Cataloging-in-Publication data

Guttman, Jon
Grim Reapers: French Escadrille 94 in World War I / by Jon Guttman.
p. cm.
ISBN **978-1-935881-40-7**
1. Crehore, Austen Ballard, 1893–1962. 2. Marinovitch, Pierre, 1898–1919 3. Martenot de Cordoux, André, 1893–1991.
4. World War, 1914–1918 --Aerial operations, French.
5. Fighter pilots -- France. 6. Aeronautics, Military --France --History. II. Title.

ND237 .S6322 2011
759.13 --dc22
2011904920

Table of Contents

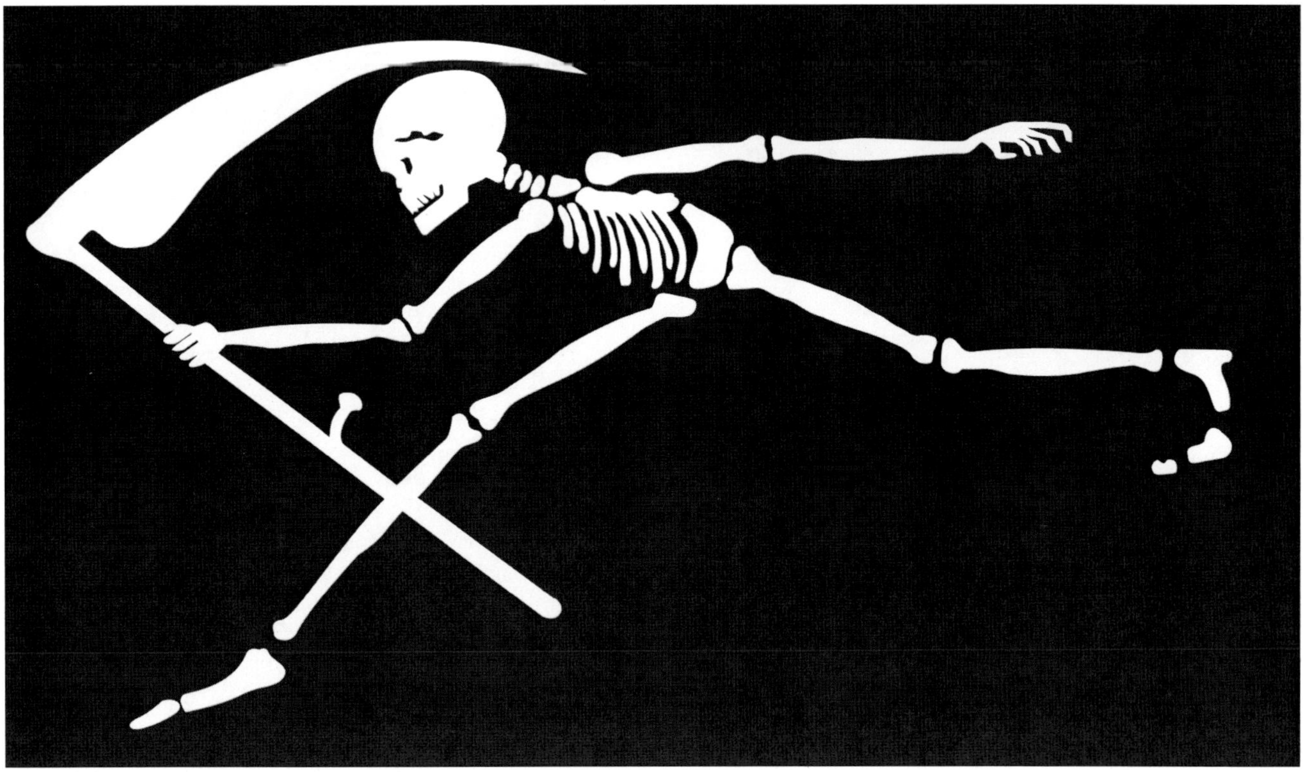

Foreword

World War I was not the first conflict to see the use of aircraft over the battlefield – that had occurred as early as June 1794, when the French used a hydrogen balloon to observe Austrian activities around Maubeuge and Fleurus. It was not even the first to see the military use of airplanes, which has previously operated during Italy's invasion of Tripoli in 1911, in the Balkan Wars of 1912 to 1913, and during the concurrent Mexican Revolution. It was, however, the first war to see a struggle for aerial dominance as serious as those taking place on land and sea. In 1915 it introduced a tool toward that end that captured the public imagination, the fighter plane, whose more successful pilots became celebrated as "aces."

As the fight for the sky escalated, lone duels between "knights of the air" gave way to confrontations between squadrons, some of which became as famous as the individuals who contributed to their reputations: units such as German Manfred *Freiherr* von Richthofen's *Jagdgeschwader* I, more widely known today as "the Red Baron's Flying Circus," France's stork-decorated *escadrille* Spa.3 "Les Cigognes," and the United States Army Air Service's "Hat in the Ring," outfit, the 94th Aero Squadron. Less successful and less famous units nevertheless contributed their share to their respective countries' causes, with achievements and deeds of their own to be chronicled. Each had its own special character and posterity has provided scholars interested in preserving their histories to varying degrees.

Above: A formal photograph of *Sous-Lieutenant* André Martenot de Cordoux at the end of the war, proudly displaying his medals and a silver pin modeled after his *Morte Qui Fauche* emblem. (Service Historique de l'Armée de l'Air, B92.4178)

The subject of this book, *escadrille* 94, was neither the highest nor lowest scoring of France's fighter squadrons, but it made a significant contribution to Allied victory and boasted several aces in its ranks, as well as several American volunteers farmed out through the Lafayette Flying Corps (LFC). One of the French aces who served in it, André Martenot de Cordoux, looked back on it this way:

"It's true that 94 didn't have the benefit, as did some other *escadrilles*, of a certain notoriety, despite all manner of services rendered to the country by its pilots. It was little-known compared with the famed '*Cigognes*.' One cannot really compare such things, one unit having already been assembled since the beginning of the war as an *escadrille* (N.3) before going on to constitute *Groupe de Combat 12, 'Les Cigognes'* under the command of Félix Antonin Brocard, in November 1916. Certainly, their publicity was well-deserved. In contrast, poor 94 didn't appear until a year later and was not endowed with a leader of Brocard's caliber.

"That said, I believe that I can claim that 94 gave a very honorable account of itself by the time of the Armistice, thanks to the teamwork of all pilots, stimulated by the combative ardor of some, as history may judge from its record."

Reviving this particular unit's exploits was made possible by adding flesh to a skeleton. The skeleton came primarily through the prodigious research efforts among the French archives by the late Frank W. Bailey, who bequeathed the fruits of his labors to me in hopes that I could add more to flesh out that framework. In the case of *escadrille* 94, some additional tissue came from the literature left behind by an American member, Houston Woodward, and by LFC chroniclers James Norman Hall and Charles Nordhoff. The most however, came through a series of interviews I had with André Martenot de Cordoux. It was Martenot's detailed memories of day to day life in the *escadrille*, as well as the occasional thrills of combat, that truly brought this creation to life. To that was added the testimony of one of the *escadrille's* less fortunate members, Pierre Chan, who became a prisoner of war, as his brother and squadron mate Henri had done a month before him.

All this, together with research into the records of some closely associated sister squadrons and the German units opposing it, has led to a detailed look at the wartime career of a fairly typical French *escadrille de chasse* that is relatively rare in the English language. As such, it adds a facet to the overall picture of the early years of air to air combat, and further insights into the highly individualistic airmen who pioneered it.

Chapter 1: Formation and First Blood

Above: A Blériot *Pingouin* on the training field at Avord. Once the fledgling airman learned to control the clipped-wing "Penguin" on the ground, he graduated to actual flight. (Henry Forster via Jon Guttman)

Escadrille N.94, as it was initially designated in reference to the Nieuport 24 sesquiplanes with which it was first equipped, was hardly the first single-seat fighter unit in France's *Aéronautique Militaire*, nor would it be the last. It was born on May 14, 1917 when the *Grand Quartier Général* (GQG) of the French Army issued *Ordre No.* 13.492, merging *Détachements de Chasse* 512, 513, and 514 at Melette aerodrome, near Châlons-sur-Marne in the IVème *Armée* sector. At the end of the year N.94 itself would be split to furnish a nucleus toward the formation of yet another fighter squadron for France's rapidly expanding air arm.

N.94's commanding officer, *Capitaine* Edouard Louis Hyppolyte Pillet, had been born on March 5, 1886. Enlisting in the army at age 28, he had entered the air service in July 1915, completing 120 reconnaissance missions as an observer until he qualified as a pilot in July 1916. In November *Lieutenant* Pillet was one of the founding members of N.79, an *escadrille corps de l'armée* (reconnaissance squadron) that flew a mixed bag of Nieuport 17 scouts and Sopwith 1 A2 two-seaters from Mesnil-St.Georges aerodrome in support of the IIIème *Armée*. Initially assigned Nieuport 17 N2095, Pillet served with the unit until May 25 1917, when he departed to take charge of N.94. During that time he logged 19 long-range reconnaissance sorties and another 12 front-line flights at low altitude between January and April, in spite of being wounded on March 21.

Besides its CO, Pillet, N.94's original complement of flying personnel were *Sous-Lieutenants* Alexandre Ottavy and Georges Parizet; *Adjutant* Pierre Desmond, *Sergents* André Martenot de Cordoux, Léon Van Den Bosche, Henri Variot, Jean Garnier, Henri Gross, Jean Régis and Marcel Moreau; *Maréchaux-des-Logis* Pierre Barriot and Léo Rayer; *Caporaux* Emile Champagnole and Pierre Marinovitch; and *Brigadier* Marcel Bostière. The *détachements de chasse* from which most of them came had been small fighter escort flights serving as adjuncts to *corps d'armée escadrilles*, providing escort for their two-seater machines during forays into enemy territory. *Détachement* 512, for example, had been attached to *escadrille* F.22, equipped with Farman F.40 pushers, 513's Nieuports had escorted the Caudron G.4s and G.6s of C.56, and 514 had done similar duty for F.71's Farmans.

Although each pilot had come to N.94 by his own particular route, it might be instructive at this juncture to look into the background of one of its principal protagonists. Joseph Marie André Martenot de Cordoux was born on March 14, 1893, in Chalezeule, Doubs, near the Swiss border, the son of 34-year-old Marie Charlotte Bathilde Martenot de Cordoux and her 24-year-old second husband, agriculturalist Claude François Jean. Attending the *École* Violet, André was excused from mandatory military service for a year due to physical weakness as of November 26, 1913, and also got a deferment toward the completion of his engineering studies.

All that became moot when Austria-Hungary declared war on Serbia on July 28, 1914. On August 12, nine days after France and Germany declared war on each other, 21-year-old Martenot enlisted in the army, a decision on which he still looked back with pride when interviewed 64 years later:

"Engaged in the infantry the day after war was declared,

Above: Martenot takes the controls of a Caudron G.3, in which he took advanced training as a reconnaissance pilot. This postwar plane bears civil registration letters. (A. Martenot de Cordoux via Jon Guttman)

then into aviation, in all I can tell you that by my initiative and personal combats I saved the lives of thousands of French combatants and civilians...which isn't a day›s work and necessitates precise explanations for the benefit of the incredulous."

Joining the 149ème *Régiment d'Infanterie* on August 13, Martenot arrived at the front in time for the critical Battle of the Marne in which, between September 3 and 10, 1914, the German drive on Paris was finally checked. During the counterattack that retook Souain, Belgium, on September 19 he earned his first regimental citation for surprising an advance German patrol with remarkable results.

"I was an accomplished rifleman before the war," Martenot said, "which may be why my company commander, *Capitaine* Altérac, thought me a bit haughty. He felt that I acted 'superior' or at least 'different' from the others.

"One morning I was out on a scouting patrol and, as we were spread out, I lost contact with the others. I came upon a village and, looking over the crest of the hill, I saw that it was just being occupied by some advancing Germans. I decided to try something. I yelled: 'Prepare to attack! Fix bayonets!' as loud as I could, hoping to make them think my whole outfit was going to descend on them. The nearest Germans said '*Kamarade*', and surrendered. Then more came out with their hands up. I couldn't believe it. My silly ruse was spreading down their ranks like a rumor among gossips, beyond my wildest hopes.

"Five of my pals turned up just then, and we collected the prisoners. More began appearing. Germans were just accumulating. It must have been a young, inexperienced unit. I found a German who spoke French and told him to tell his commanding officer that the village was surrounded by my regiment, and that he should order the rest of his men to surrender to avoid a futile slaughter. He did. In all, we collected 128 prisoners and shut them in the church, then I sent one of my comrades to fetch the captain. When he arrived, seeing me in the middle of the Germans, chatting with certain ones who spoke French, he exclaimed: 'Well, then, so it's you again!' without further word of satisfaction or compliment."

Martenot would further distinguish himself – and be wounded – during the Battle of the Yser River, during which his unit played a part in thwarting a last enemy attempt to flank the Allied armies between October 22 and November 15, 1914. Soon after that, the German drive in Flanders ground to

Above: A highly evocative aerial photograph of a Caudron G.3 on a training flight. (SHAA B81.1272)

a halt and the Western Front, stretching from the North Sea to the Swiss border, settled into the miserable stalemate of trench warfare, punctuated by costly, futile attempts by both sides to regain the initiative. Martenot described how he took a hand in foiling one such effort by the enemy on May 2, 1915:

"In the trenches 30 meters from the chapel of Lorette, I detected the presence of German miners and warned Headquarters, thus preventing our being surprised. At daybreak the next morning, May 3, the lines leapt up in the air over several kilometers as the enemy sappers' underground charges were detonated, causing the deaths of thousands of men and considerable damage. But, in part through my information, our troops succeeded in slowing the enemy advance and partially repulsing it. Four days later the enemy was to lose to a large degree the benefits of his costly attack. And during those combats no one knew the truth, the grain of sand, my information, which had upset and to a large part denied the Germans the benefits of their attack."

And so it went. Attack and counterattack with little or nothing achieved but mass, anonymous slaughter. Martenot watched his comrades die around him ("*Capitaine* Altérac fell in my arms," he recalled, "having received a bullet in the right side of his head") and desperately sought a way out of the trenches. As for many others, he turned to aviation – new, dangerous, but of tremendous unexplored potential and above all a last opportunity for individual initiative in a war that was becoming more mechanical and less human. His application for a transfer into the *Aéronautique Militaire* was approved on August 5, 1915, and he commenced flight training with the *2ème Groupe d'Aviation*.

"I started my training at the school at Pau, in 25-hp and 45-hp '*Pingouins*,' Martenot said. "The *Pingouin* was a Blériot plane with its wings cut short. It was not truly an airplane; it hopped a few meters in the air; with a 45-hp motor you could bounce about 20 meters, but not for long. You simply rolled about the field in them; the instructor or '*moniteur*' would let you go on your way across the field, then turn and come back. The *moniteur* would estimate how much control you exercised. You went strictly solo; that was to build confidence in a pilot, it simulated lone flight. We had a nice large field, so no one would be in the way.

"Then it was on to 100-hp Blériots, learning to fly in a circle. You would do a circuit of 150 kilometers during the final test for your brevet, or flying certificate. We had several accidents every week—some killed, some injured. For my own part, I never had a mishap in training. On December 26, I earned my military pilot's brevet.

"There was, by the way, another training field at Avord that used two-seater Caudrons for dual-control training for those who wanted an alternate means to the 'Blériot' method at Pau.

"After you took the test circuit, the commandant, observing the steadiness and skill of the student, would choose the pilot for his role. The pilot was then trained further on more advanced aircraft—Nieuports for scout pilots, Voisin

Above: Another French trainer, the Maurice Farman MF.11, with Houston Woodward in the cockpit. Martenot and Woodward would later serve together in *escadrille* N.94. (Houston Woodward Collection via Jack Eder)

pushers for bomber training, or the Caudron or Farman for reconnaissance training. I was assigned to the Caudrons."

After obtaining his military bevet on December 26, 1915, Martenot left Avord on February 7, 1916 for the *Groupe des Divisions d'Entrainement* (GDE) at Plessis-Belleville, where he would continue training while awaiting assignment to an operational squadron. On April 9 he was ordered to *Escadrille* C.28 at la Ferme d'Alger, between Reims and Châlons-sur-Marne, in the Champagne sector. On April 14 Martenot reported to its commander, *Capitaine* André Volmerange, and re-entered the shooting war.

Aside from a few obsolete single-engine Caudron G.3s, C.28's main equipment was the twin-engine Caudron G.4, a remarkable contraption for its time. Designed by Gaston Caudron, who was heavily influenced by Orville and Wilbur Wrights' Fliers, the G.4 was France's first twin-engine production airplane. Although its wing-mounted engines powered tractor propellers, the Caudron's tail assembly was joined to the wings by two lattice-construction booms, an arrangement more common on pusher aircraft that earned it the sobriquet of "*Gitterrumpf*" (lattice-tail) from its German opponents. Although it was slow at 82 mph, Martenot discovered that the G.4 handled surprisingly well, though few men in their right minds would have done what he and his observer-gunner, *Soldat de 2ème Classe* Maurice de Martin, did with it.

"My time with C.28 was of short duration," Martenot said, "from April 9 to May 20, 1916. In the beginning, I carried out some flights in the Caudron G.3, but very quickly a Caudron G.4 bimotor with 80-hp Gnome and Rhône rotaries was assigned me. Its agreeable qualities permitted all of one's fancies: loopings, *renversements* and glides with engines cut back or stalled. It thrilled me and I profited by my freedom during the return from the lines to the airfield, when there was no enemy in sight, to give my heart joy, so that after about 15 days of intensive training I became the best acrobatic flyer in the *escadrille*. When he was my passenger, Martin would get completely excited, and in glides without the engines on he did not hesitate to leave his place in the cockpit to sit on the upper wing and shake my hand! I was only a *caporal* at the time, and I don't think the officers with whom I carried out my regular missions over the lines appreciated my and Martin's horsing around without parachutes!

"Our usual duties were most varied: Visual reconnaissance, artillery spotting, occasional bombing with small caliber bombs, infantry liaison, machine-gunning enemy lines, etc. The observer/passenger situated ahead of the cockpit was generally a cavalry officer reclassified into aviation, the role of cavalry having proven itself, since the very start of the war in 1914, to be disastrous in the face of new weapons. The officer/observer was therefore responsible for the mission. In favorable weather, we accomplished from one to three sorties a day without any particular enthusiasm, hardly able to prove our initiative. After those regular missions, the *escadrille* commander authorized me to carry out personal or special flights with my comrade Martin, who was named '*mitrailleur*' (machine-gunner) for the occasion. It was with him that I obtained my first success on May 20, 1916."

Returning from a voluntary reconnaissance sortie, Martenot and Martin encountered two German LVG two-seaters, each mounting one fixed, synchronized machine-gun in the front and one on a flexible ring mounting for the observer. Martenot promptly attacked them. In the fight that followed, the Frenchmen came as close as four meters to one of their antagonists, who got in a burst of fire at point-blank range, putting three bullets through Martenot's right leg and foot.

"I was wounded first," Martenot said, "but nevertheless continued some minutes until he was shot down, his disorderly flight making me think he was also hit. He went into a spin on his right wing and thus spun down for about 1,500 meters. I then lost sight of him, for he became hidden in the ground mist. Not being able to recover up to then, however, he hadn't any chance to do so thereafter. Meanwhile, the other German had retired. My right engine was out of commission, but I managed to land my machine on one engine.

"As I arrived at Ferme d'Alger all personnel – officers, pilots, mechanics – were assembled before the hangars awaiting

Above: A Caudron G.4 at Issy-les-Moulineaux demonstrates the sprightly handling that inspired *Caporal* Martenot to indulge in aerobatics while flying it in C.28. He also scored his first confirmed aerial victory in a G.4. (SHAA B93.229)

my problematic return, having been alerted by the artillery batteries' observation posts near the lines that had followed our combat, which had lasted, it seemed, about 10 minutes before the German's fall. Don't forget, this was in early 1916, and combats of that sort were rare. I believe in fact that it was the *escadrille's* first. It is incontestable that the German machine crashed in the indicated region, east of Reims, confirmed by numerous witnesses from different sources. The pilot was probably wounded in the combat like myself, and perhaps killed, which would explain the plane's falling into a spin."

Such would indeed seem to have been the case. German records noted the death of *Vizefeldwebel* Albert Broemel at Warmeriville that day, although his observer seems to have survived.

Sent to the hospital for several months, Martenot received his first citation for service in the air and was promoted to *sergent* on June 22. When he was finally released to return to the air service on February 14, 1917, his right leg was seven centimeters shorter than his left as a result of his wounds. In spite of that, both he and his superiors realized that Martenot's aggressive initiative was better suited to fighters than observation planes. After further instruction in aerial gunnery and flight training in single-seaters, he was assigned on March 24 to C.56 at Ferme d'Alger. Although this was another Caudron G.4-equipped unit, Martenot spent most of his time flying a Nieuport 16 in the *escadrille's Détachement de Chasse* 513.

Sometimes operating with C.56 was N.38, based at nearby La Noblette aerodrome. Commanded by *Capitaine* Jean d'Harcourt and equipped with various Nieuport models, N.38 boasted several distinguished pilots in its ranks, including *Adjutant* Gustave Douchy and *Sous-Lieutenant* Georges Félix Madon. Douchy would eventually be credited with nine German aircraft shot down. Earlier in the war, Madon had force-landed in Switzerland, was interned and later escaped. As a member of N.38, he would become the fourth-ranking French ace with 41 victories, although it is generally believed that if one confirmed his probable claims, which could not be confirmed by strict French standards, his actual score would have been around 100.

From April 6 through June 11, 1917, Martenot was attached to N.38, which gave him the opportunity to learn some of the finer points of aerial combat from Madon, who he regarded as one of the best pilots of the time. He was not alone in either his assessment of Madon or in having what he considered the privilege of honing his skills under his tutelage.

On April 17, 1917, the Third Battle of the Champagne began and would continue until July 16, by which time the IV*ème Armée* gained the hill positions of Auberive sur Suippe, Mont Sans Nom, Mont Blond, Teton, Casque and Mont Cornillet. During the fighting, C.56's Caudrons and Nieuports were engaged in constant reconnaissance missions, often in concert with N.38.

"In preparation for an attack in the Champagne" Martenot recalled, "I went on a photographic mission in a 13-meter Nieuport with a camera mounted at my back, behind the seat, with 40 automatically-changing plates. I would fly back and forth over the sector to be photographed, exposing my plates at intervals – and exposing myself to attack by any enemy fighters that should choose to interfere. The first mission went smoothly at first, then the weather changed – too many clouds. I had to go low to take proper photos.

Above: This head-on photograph of a Caudron G.4 of *escadrille* C.212 shows the limited field of fire afforded the observer by the close-set engines and propellers. In spite of that handicap, several French aces managed to score their first victories in Caudron G.4s, including Martenot. (Imperial War Museum)

"Photography varied from day to day; sometimes good, sometimes impossible. On one occasion, I was up in my 13-meter Nieuport, marked with a red 'X' on the side of the fuselage to distinguish its duty, with *Adjutants* [Georges] Babo and [Adrien] Fauvet [of N.38] flying protection for me. We were flying along at 800 meters altitude in a 'V' formation, with me in the lead, conducting my photographic duties, when we were dived on by five German Albatros scouts. My escorts were occupied with four of the Albatros, while the head of the patrol attacked the photo plane – me! With all the advantages in the enemy's favor, there was not much for us to do but try to fight our way out, and it was a miracle that we all survived. Fauvet came down in French lines okay, but I was badly wounded in the arm and crashed, further cutting my arm, and had to return to hospital for awhile. They counted 84 bullets in Babo's plane, but he also made it back."

Martenot claimed that during his time with C.56 and N.38 he first began conceptualizing a personal marking that would eventully become the unit insignia of *escadrille* 94. "Up until now," he said, "my activity in the various units to which I was assigned had been carried out under the emblem of the *escadrilles* or just anonymously in special cases, in order to deceive the enemy. That became constricting to the crews with whom I collaborated in protection of reconnaissance machines, artillery spotting, special missions, etc., who wished to maintain recognition contact. I was thus led to adopt a personal marking. Since the clearest of our war activities was to kill one another, the idea of '*La Mort Qui Fauche*' ('Death Who Mows' – the Grim Reaper) came to mind.

"But how to realize it? The usual position of the reaper is to hold his scythe to the contour of the ground or, in his travels, on the shoulder, facing backwards; neither of those seemed to be worthwhile poses for my subject. I therefore opted for the position of his holding it before him like a flag, giving a more dashing, unrestrained manner, in my opinion more satisfactory for pursuit; in exaggerating the length of his members I obtained, in my opinion, an impression of speed, of haste in his evil designs; and then, the idea of so impressing the enemy didn't displease me!

"I didn't have time to occupy myself with realizing those ideas on the side of my plane, but for the time being I confided them in a drawing to a comrade going to Paris, and it was a jeweler on the *Avenue de l'Opéra* who realized the model in silver, to be carried on my tunic. It was identical to the emblem that had become something of a special infatuation to me, which I planned to reproduce, as large as possible, on either side of my airplane's fuselage as soon as opportunity permitted.

"I was not the only one who entertained such notions. In N.38, Madon had his machine painted entirely red, so it would be recognizable from afar, while Douchy painted his

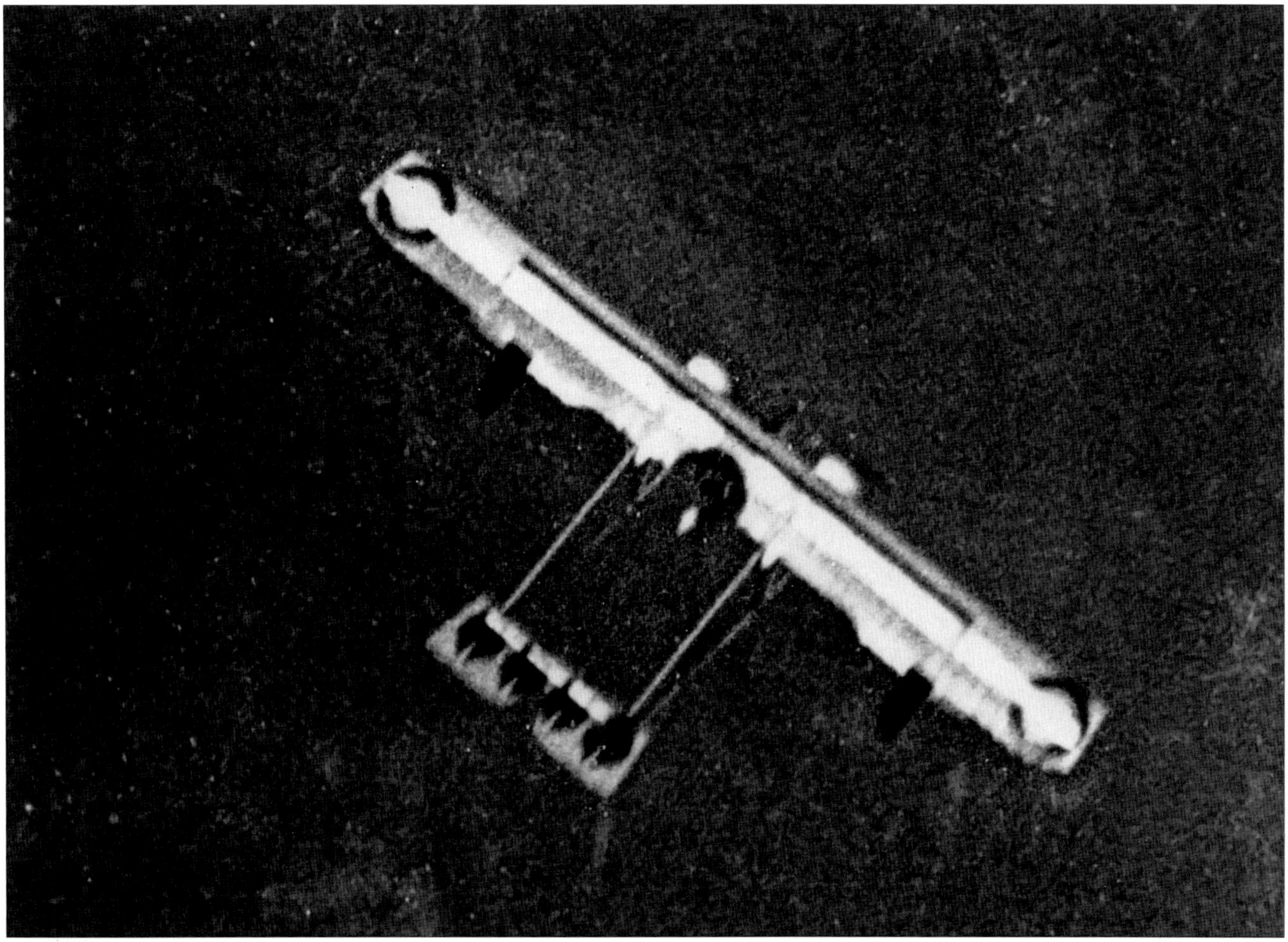

Above: An in-flight photograph of a Caudron G.4 of C.30 on a reconnaissance mission. (SHAA B92.1716)

Spad mostly blue. In any event, I was not to paint my creation on my own plane before I was transferred, one last time, to a newly-formed *escadrille de chasse*. N.94 was actually formed on June 1, 1917, when it partially occupied Melette near Châlons-sur-Marne along with N.38, and in the first few weeks of June and July, while being assigned to N.94, I continued to carry out special missions for my old squadron over the same territory, since upon formation 94 had reduced activity."

There was enough activity attending N.94's formation for one of its members to score its first victory early on. On May 25 *Adjutant* Pierre Desmond, formally of F.71's *Détachement* 514, shot down an enemy plane, for which he was cited on June 18. His victim may have been *Leutnant der Reserve* (*Ltn.d.R.*) Alfons Paulus of *Jagdstaffel* (*Jasta*) 21, who was killed over Dontrien that day, although he was also claimed by French anti-aircraft gunners. N.94 also suffered its first loss when *Sergent* Henri Gross failed to return from a patrol, his Nieuport possibly being the one claimed over nearby St. Hilaire le Petit by *Ltn.d.R.* Rudolf Matthaei of *Jasta* 21. German records also noted the capture that day of Nieuport 23 No.2920 whose pilot, a "Corporal LeGras," was taken prisoner.

With the *escadrille* blooded in battle, Martenot said, "One of the preoccupations of the new 94 was naturally to find an insignia. The first training flights and operational sorties were carried out with planes carrying no distinctive markings. After some days of waiting, *Capitaine* Pillet, our new CO, proposed his idea to us: a crab. When he contacted me a little while afterwards for comment, I uttered a cry of horror, telling him that if he persisted in those views I'd prefer to ask for a change in assignment. Surprised at my reaction, pursuing his idea, he sought to convince me, remarking that the said animal had huge pincers. 'Yes,' I told him, 'but you forget that when faced with danger the crab buries himself in the sand, draws back and hides – a lovely example for pursuit pilots.'"

In spite of Martenot's negative reaction, the crab was adopted in August 1917 and appeared on N.94's Nieuports until December. It was then augmented by an added light blue disk in which the female personification of "*la République*" appeared. Throughout those months, however, Martenot persistently petitioned to replace the crab.

"There remained the matter of personal identity," Martenot added. "From one *escadrille* to another, the use of insignias was different, following the whims of their commanders or certain pilots. Besides the common unit insignia in 94, we had Arabic type numerals, very important and useful for recognition. I'm not superstitious and had no preconceived idea as to my choice;

Above: Caudron G.6 with the winged scarab emblem of C.56, a unit in which *Sergent* Martenot briefly served. Although more advanced looking than the G.4, the G.6 offered no significant advantage in performance, and did not handle as well. (SHAA B75.622)

it turned out that I had the number 5. I don't remember the reason for that assignment, but I had no desire to change the decision. *Capitaine* Pillet, who did not fly, had no number; others certainly had them, but I don't remember them, save, for some reason, the number 16, which belonged to Miguel Sainz."

N.94's Nieuport 24s and 27s made their first sortie over the lines in squadron strength on June 23. The patrol was uneventful. A patrol on July 13, however, was, as *Sergent* Henri Variot sent an Albatros D.V crashing in German lines near Mont Haut. The Germans noted the death of *Vizefeldwebel* Wilhelm Neuhaus at Beine, but further information on this airman, including whether or not he died in a single-seater, is lacking.

Born in Paris on November 14, 1897, Henri Variot was the second son of the chief physician at the Trousseau hospital. After completing classical studies and earning two baccalauréats at the Lycée Carnot, he enlisted in the air service on July 15, 1915, and after entering flight school at Étampes in January 1916, soloed in a Farman in March, and received military brevet No.3740 on July 6. Promoted to *sergent* on August 15, he was assigned to F.22, flying low-level reconnaissance and close support missions over Verdun and the Somme. On August 31 he got his first citation:

"He has revealed himself since his arrival at the *escadrille* as a pilot of the first order through his boldness as well as for audacity and composure. On August 9 he flew at 350 meters to maintain liaison with the infantry in spite of enemy machine guns. His motor was badly hit, but thanks to his calm and energy he landed near the French batteries."[1]

Variot was dissatisfied with the slow, lumbering Farmans he had to fly and at the end of 1916 his request for fighter training bore fruit when he was sent to Pau to train on Nieuports. Upon completing the course in January 1917, he was assigned on February 12 to *Détachement* N.512 of *Escadrille* F.22, flying Nieuports in support of his old unit and occasionally flying patrols with N.38. Then, on June 1, he joined N.94.

On July 18, N.94 got a new executive officer (XO) in the person of *Lieutenant* Henri Constans de Saint-Sauveur. Born at Saint Sauveur Gaillac on September 8, 1887, Henri Marie Saturnin Constans de Saint-Sauveur entered military service on September 28, 1908, with the 19*ème Régiment de Dragons*. Rising to *brigadier* on February 6, 1909 and to *maréchal-des-logis* on December 9, 1910, he underwent some additional training to attain his *sous-lieutenant*'s commission in the *Réserve* on September 28, 1911. After having been cited for heroism on November 20, 1914, de Saint-Saveur decided to leave the dragoons for the air service, as an observer in Bl.18 on December 1, and in Bl.9 on the 25th. On February 2, 1915, he began pilot training, earning French Aero Club Pilot Brevet No. 1825 on April 1, and Military Pilot Brevet No. 1325 on the 20th. Assigned to MF.54 on May 10, he was promoted to *lieutenant* on September 20.

De Saint-Saveur entered the high-stakes realm of the fighter pilot on February 25, 1916, when he was given command of N.67. His transfer to N.94 as second in command seems to have been something of a refresher, to hone his skills at

Above: A mechanic of N.38 chats with *Adjutant* Georges Madon, seated on the wheel of his red-nosed Spad 7, while *Adjutant* Gustave Douchy looks on. Officially credited with 41 victories, Madon was also the mentor to other aces, including André Martenot de Cordoux and David Putnam. Douchy, Spa.38's second-ranking ace with nine victories, often flew a blue Spad to identify himself in formation. (SHAA B92.32)

Below: French airmen on the Champagne front between March 28 and April 18, 1918 include (from left) *Sous-Lieutenant* Georges Babo (MS.156), *Lieutenant* Georges Madon, commander of Spa.38, and *Sous-Lieutenants* Jean Casale and François Brun, both of MS.156. In the background are a Breguet 14 A2, a Nieuport 27 and what seems to be Madon's Spad XIII S730. (SHD/A)

Above: Manfred *Freiherr* von Richthofen did not have a monopoly on red aircraft. *Sous-Lieutenant* Madon had the fuselage, undercarriage and vertical stabilizer of Spad VII S1451 painted red for easy identification when N.38 was on patrol. He later did the same with at least two Spad XIIIs and a cannon-armed Spad XII, S434. (SHAA B75.363)

both aerial combat and leadership in preparation for a new command, which he would receive at the end of the year.

On July 25 Martenot, who had been promoted to *adjutant* on the first of the month, was flying a lone patrol in his Nieuport 24 when he encountered a Rumpler two-seater over Mont Cornillet, east of Reims. "I remembered what Madon had told me about how to attack a biplace," he said. "Keep out of the line of fire of his rear gun. I approached from slightly below, zig-zagging from side to side until I was ready to take aim, then I fired from behind and below."

Martenot's bullets struck home and the Rumpler went down to crash, killing *Gefreiter* Bruno Jakob and *Leutnant* Heinrich Burkert of *Flieger Abteilung (Artillerie)* 206. The victory was confirmed as his second. The month's only sour note was struck on the 28th, when *Sous-Lieutenant* Ottavy's engine gave out as he was landing—due, it turned out, to faulty gears. His Nieuport fell 30 meters to crash on the field, but the former teacher from Toulon had reason to count his blessings—although he suffered injuries around one eye, the eye itself was unharmed.

On September 5, N.94 received its first Spad VII, serial number S135, which was issued to Martenot on the 9th. He quickly took to the new plane and on September 11, he used it to bring down a German two-seater west of Beine, though that plane's demise was not confirmed. The rest of the *escadrille* continued to fly Nieuports until early 1918 when, entirely re-equipped with Spads, it was re-designated Spa.94.

The Spad VII with its 180-hp direct-drive Hispano-Suiza 8Aa engine became Martenot's favorite fighter. Less maneuverable than the Nieuport 24, it was nonetheless nimble enough to hold its own against its German opponents in a dogfight if necessary. In a diving, hit-and-run attack, there were few planes comparable, largely thanks to the Spad's robust wing structure, which could hold up to the stress of a power dive better than most of its contemporaries. When Martenot was issued an upgraded Spad XIII in the spring of 1918, with double the armament and a 220-hp geared Hispano 8B engine, he chose to keep his old Spad VII to fly as well:

"I would fly regular missions in the Spad XIII in order to keep formation, for the standard Spad XIII was faster than the VII. When all Spads returned, I would use my VII between patrols, on my own individual flights. Despite the single gun and lower speed, it was more maneuverable than the XIII and, frankly, the closeness of our combats rendered speed less important.

"My Spads were originally fitted with the Aldis sight for aiming the gun or guns, but it presented a problem: It distorted distance, giving the impression that the target was closer than it was. I replaced the Aldis with a triple bead and 'V' sight of my own invention, lining up my target with a set of cross

Above: Although one of N.94's founding members, *MdL* Pierre Barriot is shown in the cockpit of a Nieuport 24bis marked with the crow or raven emblem of neighboring N.155. On December 1, 1917 Barriot became one of the founding cadre of N.156, with which unit he scored his first victory with *Caporal* David Putnam on January 27, 1918. (David Méchin)

hairs without glass nearest to my eye, a bead midway down the gun, and a 'V' aperture near the muzzle. Because of this sight my accuracy became so deadly that Germans seemed to avoid combat with me."

On September 8, four aircraft of N.94 crossed the lines and became embroiled in a fight with German Albatros D.Vs, during which *Sous-Lieutenant* Georges Parizet and *MdL* Pierre Marinovitch teamed up to send one of the enemy fighters crashing to earth near St. Hilaire-le-Petit. It was credited as the first victory for both Parizet, who returned with several bullet holes in his plane, and for Marinovitch, who was subsequently cited for the *Croix de Guerre* with palm.

This success inaugurated the career of one of World War I's leading fighter aces. Born in Paris on August 1, 1898, Pierre Marinovitch was, according to Martenot, "of part Serbian extraction," with a Polish mother.

There was, in fact, more to Pierre's background than that. His grandfather, Paris-educated Jovan Marinovic, had served as the Serbian prime minster from 1873 to 1874, and as Serbian ambassador to France from 1879 to 1889. Marrying Persida Anastasijevic, daughter of wealthy merchant Misa Anastasijevic, Jovan had two sons and a daughter. One of the former, Velizar Marinovic, married Polish Agripina Bronkov and then moved to France, where they Gallicized their names as Bélisaire and Agrippine (née de Bronikoff) Marinovitch. Their son Petar – Gallicized to Pierre – was still a student when war broke out, but on February 12, 1916 he enlisted in the *27ème Régiment de Dragons*. He soon became disillusioned with being a cavalryman in a trench-bound infantryman's war, and on July 14 he transferred into aviation.

On September 9, 1916, Marinovitch commenced training on Farmans at Chartres, earning military brevet No. 4910 on November 15, and Civil Aero Certificate No. 5181 on January 31, 1917. He then trained in aerobatics at Pau and aerial gunnery at Cazaux before going on to GDE, where he got his squadron assignment to N.38 on March 19. There he, like André Martenot de Cordoux and David Putnam, undoubtedly learned some invaluable pointers from Georges Madon, but not long after his arrival *Caporal* Marinovitch fell seriously ill and spent two months in the hospital. Upon recovery, he learned

DÉCORÉ PAR LE GÉNÉRAL GOURAUD
Le général Gouraud, venu décorer des aviateurs, parmi lesquels le grand as Madon, remet une palme à l'adjudant Henri Variot, qui fut cité quatre fois à l'ordre du jour.

LES HÉROS DISPARUS : HENRI VARIOT

L'AUTRE semaine nous avons conté la vie et la fin glorieuse de l'adjudant Triboulet, fils de l'éminent docteur. C'est encore d'un enfant de grand médecin que nous parlerons aujourd'hui, de l'adjudant Henri Variot.

Ce jeune héros était né à Paris le 14 novembre 1897. Il était le second fils du docteur Variot, alors médecin chef de l'hôpital Trousseau. Après avoir terminé ses études classiques au lycée Carnot et passé ses deux baccalauréats, il s'engagea dans l'aviation, le 15 juillet 1915.

Il reste plusieurs mois au camp de Longvic, près de Dijon, et dès le début de 1916, est envoyé comme élève pilote à l'école d'Etampes où il reçoit les conseils du chef de piste Gougenheim qui fut son premier maître et dont il appréciait particulièrement l'enseignement. Instruit sur Farman, Variot obtient son brevet militaire en mars 1916, après un apprentissage rapide et brillant.

Il est envoyé au groupe des divisions d'entraînement où il attend d'être affecté à une escadrille du front.

Le moment arrive bientôt, accueilli avec joie, et Variot part dans la région de Verdun.

Après Verdun, la Somme. En août 1916, le jeune pilote se signale par ses nombreux vols de liaison d'infanterie à très basse altitude. Ils lui valent le 31 août une première citation :

« S'est révélé dès son arrivée à l'escadrille comme un pilote de premier ordre, tant par sa hardiesse que par son audace et son sang-froid. Le 9 août 1916, a volé à 350 mètres pour assurer la liaison avec l'infanterie, malgré le feu des mitrailleuses ennemies. A eu son moteur gravement atteint, a pu grâce à son calme et à son énergie, atterrir à proximité des batteries françaises. »

Malgré les émotions et les incidents dramatiques de ces sortes de mission, Variot ne trouve pas assez mouvementé le travail qui lui est assigné. Les gros appareils, lents et peu maniables ne sauraient l'intéresser. Il leur préfère les petits engins de combat fringants et souples. A la fin de 1916, il obtient, après quelles difficultés, de voir réaliser son rêve. Il est affecté au Nieuport. Il est envoyé à l'école de perfectionnement de Pau où un stage de courte durée suffit et en janvier 1917, il part, comme sergent, au groupe de chasse de la e Armée.

C'est là qu'il obtint deux nouvelles citations et qu'il reçut les galons d'adjudant. Il abat deux avions et fait preuve d'une rare énergie. Très aimé de tous ses camarades qui l'admirent, il n'est pas favorisé par la chance. Sa valeur, son courage mériteraient plus de succès, mais au cours des multiples combats qu'il livre il épuise la gamme de toutes les difficultés, de tous les soucis qui peuvent provenir du moteur ou de l'armement... et de l'adversaire qui fuit, désemparé, mais qu'on ne peut faire homologuer. D'ailleurs, Variot ne chasse pas pour augmenter sa totalisation : tuer du Boche est son seul désir. Et combien ont dû s'écraser au sol qui y avaient été expédiés par ce jeune, plein d'entrain et d'audace

Le 27 juillet, il reçoit une citation :

« Pilote d'une conscience élevée, a toujours accompli ses missions en homme de devoir. Le 13 juillet 1917, au cours de l'une d'elles, a abattu un avion ennemi dans ses lignes. »

Puis, le 6 novembre 1917 :

« Pilote d'une froide bravoure, possédant à un haut degré la notion du devoir ; toujours égal à lui-même dans les différentes missions qui lui ont été confiées, a attaqué, le 1er novembre 1917, deux avions ennemis croisant dans leurs lignes et a abattu l'un d'eux en flammes. »

Ces succès le font proposer quelques semaines après pour le grade de sous-lieutenant. Mais Variot voudrait obtenir la Médaille Militaire avant de recevoir les galons d'officier. Nulle mission n'est trop périlleuse pour lui.

Au début de 1918, il est affecté à une autre escadrille. Il n'y restera pas longtemps. Le 19 janvier, au cours d'une patrouille, il rencontre six Boches. Il n'hésite pas, seul contre tous, il engage le combat, mais la partie est par trop inégale. Il lutte avec une énergie admirable. Il pourrait chercher le salut dans la fuite Il n'y songe même pas et succombe sous l'attaque de ceux qui ne se sentent forts que lorsqu'ils sont en nombre. Variot avait vingt ans et deux mois.

Voici sa dernière citation :

« Pilote d'une conscience élevée et d'une bravoure à toute épreuve, chasseur remarquable déjà cité trois fois. Le 19 janvier 1918, seul contre six avions ennemis a soutenu un combat dans les lignes ennemies et n'a succombé qu'après une lutte acharnée. »

On ne savait quel avait été le sort du malheureux, porté seulement « disparu ». Hélas ! il n'était pas prisonnier et, le 25 février, ses parents recevaient l'avis de sa mort par une lettre de la Croix-Rouge de Genève, contenant les renseignements suivants, communiqués par Berlin : « Le pilote Henri Variot a été abattu en flammes, le 19 janvier 1918, à 4 h. 10 de l'après-midi à l'est de Tahure. Il était mort sur le coup. Le lieu de sa sépulture est Dormoiseral, à l'est de Tahure. »

J. M.

L'ADJUDANT HENRI VARIOT
Spécialiste des liaisons d'infanterie à basse altitude, vainqueur d'un avion le 13 juillet 1917 et d'un autre le 1er novembre 1917, tué le 19 janvier 1918, en attaquant six Boches.

Above: A rare photograph of a Nieuport 24 of N.94, with the crab insignia originally created by the commander, *Capitaine* Edouard Pillet. *Sous-Lieutenant* Alexandre Ottavy suffered facial injuries in this rough landing at Melette on July 28, 1917. (SHAA B87.4875)

that he was reassigned to the newly formed N.94. There, he was promoted to *maréchal-des-logis* on July 26.

According to Martenot, Marinovitch had been "educated in Paris and in Ireland, and spoke fluent English, which was to stand him in good stead with our later American *escadrille* members. Marinovitch preferred attacking enemy aircraft and getting citations to other missions...and he was very good at it, as his 22 victories testified. 'Marino' was a good shot, but not a good pilot. He wrecked a lot of planes. He used no tactics—just charged in and relied on his good shooting.

"Once," Martenot concluded, "he landed too low at the edge of the field, where there was a road bordered by a ridge. He bounced on the ridge and cracked up—that was a typical Marinovitch landing."

On September 13, N.94 was ordered to cease operations because structural defects had been found in the wings of its Nieuports and they had to be withdrawn for modification and strengthening. The Nieuports were returned on October 9, but 11 days later the squadron lost a member to other causes. While ferrying over a new Nieuport 27, *Sergent* Jean Régis lost control of his plane in a sudden gust of wind and crashed. He died several hours later in the hospital at Auxerre.

On November 1, Variot—who had been promoted to *adjutant* on September 7—reopened N.94's account when he intercepted two enemy two-seaters crossing the lines and shot one down in flames over Souain. His likely victims, *Leutnante* Friedrich Hahn and Karl Mangold of *Fl.Abt.* 247, were killed. On the following day, Martenot was awarded the *Médaille Militaire*. Variot was being considered for a *sous-lieutenant*'s commission, but he felt unworthy of the honor until he had earned a *Médaille Militaire*. Until then, he was willing to take on any mission.[2]

December 1 saw the arrival of new personnel to N.94—*Maréchal-des-Logis* Miguel Sainz, *Caporaux* André Martin and Charles Nicholas, and *Brigadier* Eloi Bonnet. Two days later, the *escadrille* got its first two Americans via the Foreign Legion and the Lafayette Flying Corps. Expanded from the original organization that had founded France's already famous all-American *escadrille*, N.124 *"Lafayette,"* the LFC farmed additional volunteers in small parcels to various other French squadrons. The first of them, Wallace Charles Winter Jr., was born in St. Paul, Minnesota, on May 4, 1896. When he was eight years old, Wallace's father moved to Chicago, Illinois, where he attended school before being sent to the Hotchkiss School in Lakeville, Connecticut, and to Yale University's Sheffield Scientific School. Winter was in the U.S. Naval Reserve when the United States declared war on Germany on April 6, 1917, but impatience to get to the front drove him to

Facing Page: *Sergent* Henri Variot, seen beside his Nieuport, scored N.94's first victory over Mont Haut on July 13, 1917, establishing a reputation as one of the escadrille's most *aggressive* pilots. (*La Guerre Aérienne Illustrée No.73*)

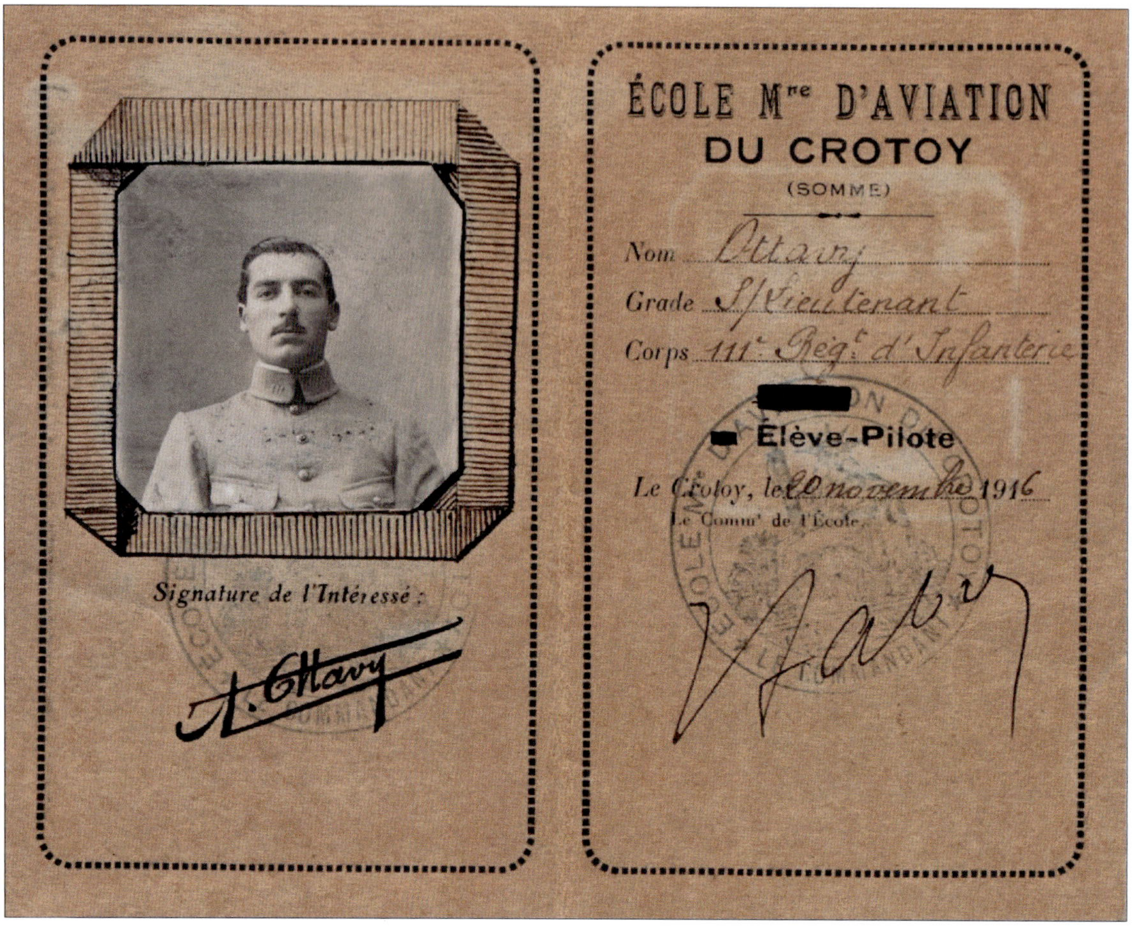

ÉCOLE M^re D'AVIATION
DU CROTOY
(SOMME)

Nom Ottavy
Grade S/Lieutenant
Corps 111e Régt d'Infanterie
[redacted]
[redacted] Élève-Pilote
Le Crotoy, le 20 novembre 1916
Le Comm^t de l'École

Signature de l'Intéressé : A. Ottavy

Above: *Sous-Lieutenant* Alexandre Ottavy's identification card. (Christophe Cony Collection)

Above: The crab insignia used by Spa.94.

Above: *La République* insignia used by Spa.94.

Above: *Sous-Lieutenant* Alexandre Ottavy photographed in a Nieuport; part of the Spa.94 crab insignia is just visible at the lower right. (Christophe Cony Collection)

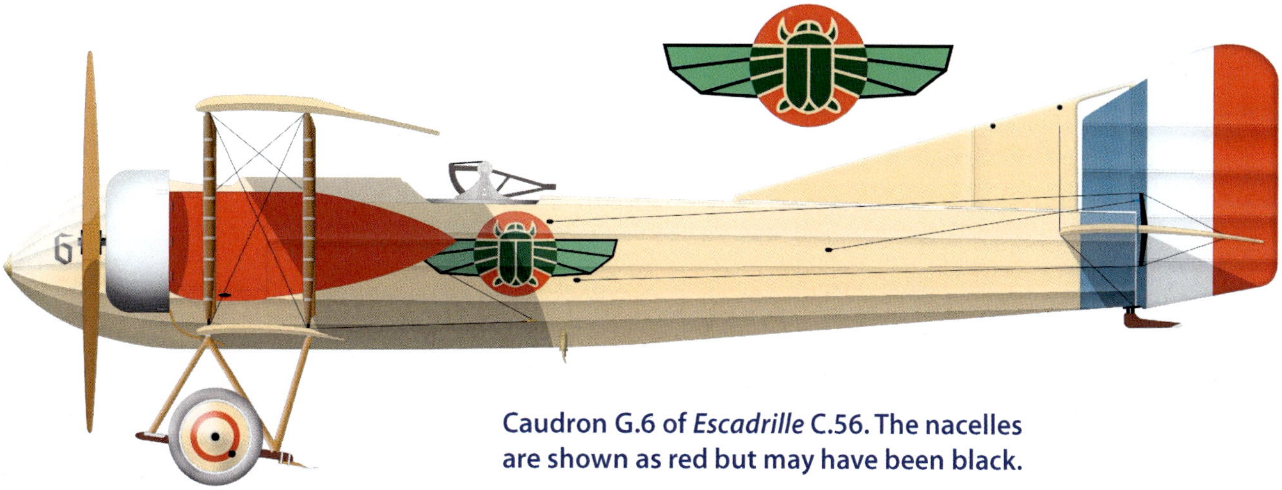

Caudron G.6 of *Escadrille* C.56. The nacelles are shown as red but may have been black.

Left: *Sous-Lieutenant* Georges Parizet in the cockpit of Nieuport 24 No.3. He apparently named his plane, but only the letter "M" can be made out below and in front of the cockpit. His first victory, on September 8, 1918, was shared with *Maréchal-des-Logis* Pierre Marinovitch, the first for both airmen and the third for N.94. (SHAA)

request an honorable discharge from the USNR in order to join the Lafayette Flying Corps. He left Yale in May and steamed to France, where he commenced training at Avord on June 29. After earning his brevet in a Caudron on September 15, and undergoing further training at Pau and the GDE, he was assigned to N.94 on December 3.

Arriving on the same day as Winter was the American destined to stay with *escadrille* 94 the longest. Born in Hackensack, New Jersey on January 9, 1893, Austen Ballard Crehore was the third of nine children born to William Williams and Anna (Ballard) Crehore. His father was a civil engineer credited with innovative structural designs for bridges and buildings, including New York City's first skyscraper, on 150 Nassau Street. When Austen was seven his family moved to a farm in Elizabeth, from which he walked to a small school. The Crehores had moved to Metuchen by 1908, but soon afterward Austen encouraged his family to move on to Westfield because its schools would offer him and his younger siblings a better education.

In 1909 Austen Crehore became the first employee of the newly incorporated Merchant Fire Assurance Company. When the United States declared war on Germany on April 6, 1917, he promptly applied to the U.S. Army in hopes of joining the aviation section of the Signal Corps, but the American Examining Board rejected him due to defective hearing. He then tried to join the U.S. Navy, only to be rebuffed again. "He told me that the thinking at the time was that he would not be able to know if he were flying right-side up or upside-down," recalled his daughter, June Gulick in 2015. "Not one to take 'no' for an answer, he learned of a person visiting from France and went to the brownstone house in New York City where the man was staying. He told me how he sat on the front steps all one evening until this man returned. They talked and my father was accepted as a volunteer in the French Aviation, after which he sailed for France.

"My father was not only deaf, but he knew no French," Mrs. Gulick continued. "This didn't stop him; he was persistent, determined, and fearless. He was a team player: always going the added mile, always looking out for the other guy, and always insisting that it be fair no matter the odds against him. He told me about checking the windsocks at all the airfields in order to be sure which way to go to get back to his camp."

In France Crehore enlisted in the *Aéronautique Militaire* on July 16. Commencing training at Avord on the 21st, he managed to obtain brevet No. 8983 at Tours on September 28, and after further training at Pau he went to GDE on November 17 and was promoted to *caporal* on December 2. He was assigned to N.94 the following day.

On December 5 Marinovitch sent a Rumpler crashing into French lines near Mourmelon-Baconnes, where its crew was allegedly taken prisoner, while Van Den Bosche crashed an Albatros two-seater near Somme-Py. On the 12th, *MdL* Barriot

Above: *Adjutant* Martenot de Cordoux in a Spad VII. Issued the first such plane delivered to N.94, S135, on September 5, 1917, he later discarded the Aldis sight and devised one of his own. The round insignia aft of the cockpit is a blue disc contraining a black bust of "La République," which was combined with the crab in December 1917 and used at least until mid-May 1918, considerably longer than Martenot implied in his recollections. (A. Martenot de Cordoux via Jon Guttman)

was promoted to *adjutant* and another LFC volunteer, *Caporal* David Putnam, reported to the *escadrille*.

Born in Jamaica Plain, Massachusetts, on December 10, 1898, David Endicott Putnam was a descendant of Revolutionary War major general Israel Putnam. A gifted student, Dave graduated Newton High School to be accepted in the Harvard Class of 1920, where he became class president, class orator and an active athlete in four sports. While still in his freshman year, however, Putnam abandoned it all to enlist when the United States entered the war. Although he passed the U.S. Army's Aviation Service examination, Putnam was hopelessly below the 21-year-old minimum age requirement, so like other young men who had encountered that obstacle, he traveled to France, joined the Foreign Legion and directly transferred to the LFC on May 31. He obtained his flying brevet on the Caudron at Avord on October 17, and after further training at Pau and the GDE, he was assigned to N.94 on December 12—just two days after his 19th birthday.

Four days later, N.94 got its fourth American, 24-year-old Walter John Shaffer from Dauphin, Pennsylvania. After graduating from Dauphin High School in 1909, Shaffer had worked in the printing trade, but also had learned to fly a pusher floatplane at Essington in 1915. When the United States entered the war, he borrowed money from Lewis Gross, his fiancée's father, to book passage to France and enlisted in the *Aéronautique Militaire* on August 1, 1917. Training in Caudrons at Avord and Tours to earn his brevet on October 7, he went on to aerobatic training at Pau and a short time in GDE before arriving at N.94.

On the following day, *Caporal* Henry Howard Houston Woodward reported for duty. Born in Chestnut Hill, Pennsylvania, on February 27, 1896, Houston Woodward was the son of Dr. George and Gertrude Woodward. After attending preparatory schools at Chestnut Hill Academy and the Taft School in Connecticut, he enrolled in the Sheffield Scientific School of Yale University. He also joined the Yale Battery at Tobyhanna, Pennsylvania on June 1916, but during his sophomore year he resigned from Yale to serve in the American Ambulance Field Service. Departing for France aboard the liner SS *Chicago* on February 19, 1917, he was assigned to Section 13 and drove ambulances for a French division during the Champagne Offensive. On July 24,

Above: A camouflaged Spad VII *escadrille* 94, probably photographed in the late winter of 1918, displays a variation on the crab emblem, in black, outlined in yellow, on a red disc with "*La République*" above it. (SHAA B92.892)

however, "Woody" did what a good many of his colleagues did – keen to play a more active role in the Allied war effort, he transferred to aviation. Attending flight schools at Avord, Juvisy and Pau, he received his brevet in a Caudron on October 1.

On December 22 Marinovitch drove a Rumpler down to crash north of Pont Faverger after a difficult fight. On the 27th, *Maréchaux-des-Logis* Rayer and Grangeron, and *Sergent* Van Den Bosche flew escort for a corps reconnaissance plane, but tragedy attended their return when Léo Rayer suddenly fell into a spin and crashed to his death. Van Den Boche also crashed and was injured.

As events of 1917 reached their conclusion, Martenot recalled the squadron mates whose acquaintance he made at that time:

"The *escadrille* personnel comprised two categories: pilots of all grades and of all origins (volunteers), and those whose functions were ground maintenance (non-flying) – staff and mechanics, armorers, etc. On the ground the organization comprised, in general, a line of hangars made of easily collapsible wood frames, covered with strong canvas, called 'Bessoneaux,' to protect our airplanes, and 'Adrian' barracks of wood, equally collapsible for transport, to separately lodge the pilot officers and *sous-officiers* [non-commissioned officers] in small rooms for two, then the mechanics in dormitories and finally the service annexes: office, armory, etc. *Sergent* Henri Variot, the son of a doctor in Paris, was my first roommate in the Adrian barracks at 94.

"In the daytime, between flights, life passed on the ground close to our machines, with good terms between pilots and mechanics. Whatever the practices between ranks elsewhere, at 94 we would all meet and eat together at the same table—pilots and ground crew, officers and enlisted men alike.

"Our first commander was *Capitaine* Edouard Pillet, who led us until April 30, 1918. Pillet was not a career officer. Before the war, he'd sold men's underwear near the Porte Maillot in Paris. He was friendly and very correct with everyone, but he seemed totally ignorant of mechanics, of fighting and of combat, as well as flying. I never saw him carry out the least fighting activity."

However much of a desk pilot Pillet seemed to be in at N.94 in Martenot's eyes, his previous service record had, in fact, been a good one; as an observer, he had been cited for carrying out 120 reconnaissance missions before becoming a pilot and being given command of N.94. Martenot went on to describe Pillet's successors:

"The second commander, who succeeded Pillet, was *Lieutenant* Guy de la Rochefordière who, contrary to Pillet, seemed to want to interest himself in the combats of his pilots and joined them in their patrols. Unfortunately, inexperienced and hard-pressed, he was shot down in flames by two Fokker D.VIIs near the lines during his third patrol, on June 11, 1918.

"*Sous-Lieutenant* Benjamin Bozon-Verduraz, promoted to *Lieutenant* on July 18, 1918, came from the Stork Group's *Escadrille* Spa.3, in which he had been on the last patrol carried

Above: Spa.94 flying personnel in early 1918. Second from left is *Sergent* Henri Grimouille; third; *Adjutant* Pierre Marinovitch, fourth, *Adjutant* Jean Joseph Ondet; fifth, *Maréchal des Logis* Miguel Sainz; seventh, *Caporal* Austen B. Crehore. (A. Martenot de Cordoux via Jon Guttman)

out by *Capitaine* Georges Guynemer on September 11, 1917, in which the latter was shot down. Bozon-Verduraz had already been credited with seven official victories, two of them shared with de la Rochefordiére, upon his arrival at 94, where he obtained his ninth, tenth and eleventh confirmations on August 17, October 3 and October 9, 1918. He was the only one of 94's commanders to have gained aerial victories while with the unit, and it is regrettable that he was so late in arriving, with so little time, about five months. He stayed with us from July 8 to December 15, 1918, at the latter date sick in hospital.

"I remember some of the others I flew with," Martenot continued, "Miguel Sainz who, despite the Spanish name, was a born Frenchman to the core. So were the Jewish Chan brothers, of whom Pierre was the youngest in my patrol. Also very young was Pierre Marinovitch, destined to be our own leading ace.

"Sainz was French, but there was some evidence of his Spanish ancestry in him – he was very nervous and a bit crazy. Henri Grimouille was calm, the opposite of Sainz. But both were good fighters, and were both credited with four victories by the end of the war. Grimouille flew a lot with Marinovitch. Joseph Ondet was a normal sort, a good *sous-officier*. *Maréchal-des-Logis* Roger Busch was reserved, not having very much dash. Léon Van Den Bosche had been a miner from northern France.

"Ondet...Busch...Crehore...Woodward...good comrades all... all are gone now.

"It was amusing, Martenot continued, "to compare my memories of my comrades with the descriptions recorded in the diary of one of our Americans, *Caporal* Houston Woodward. Except for 'Marino,' he only gave their initials, but I identified them right off as follows:

"'Marinovitch. The ace of the *escadrille*. My best friend, and one of the whitest, cleanest little fellows I have ever met. Father a Serbian, mother a Pole, was born in France, educated in Ireland, has been a French citizen for several years, talks English, French, German perfectly, and Lord knows how many other languages he has a smattering of. Has the cutest pet in the crowd, a cross between a collie and a monkey, it looks like. Superb flyer. Has brought down three Boches since early December, and was the first Frenchman to get a Boche in 1918, bringing one down inside their lines New Year's afternoon early.

Above: Austen Ballard Crehore poses proudly in his new French uniform, soon to be decorated with wings, rank chevrons and decorations…if he can survive. (June Crehore Gulick)

"'P— is a very nice, lively, sporty, quick chap. He talks fair English, and is very agreeable and handsome. (*Sous-Lieutenant* Georges Parizet).

"'B— I don't care for. He's some peasant or farmer and has never overcome his original disadvantages. Hair awfully long, filthy fingernails, drinks gallons of Pinard, and yells when he talks (*Brigadier* Eloi Bonnet).

"'O— is a corker. Nice, quiet, refined little Frenchman, very modest, excellent flyer, lots of nerve. Can trust him in a scrap. Very good natured and likable. (That's Joseph Ondet, all right!).

"'De L— is one of the FFV [Féderation Française de Voile, France's international sailing club] and a prince of a fellow. Used to play golf with him at Pau. Taller than I am, very dark, with beautiful blue eyes. Has had three brothers killed and is the sole survivor of his generation. (*Lieutenant* François de Lagerie).

"'M— Pleasant, quiet, retiring former Zouave. Speaks English some, don't know him well. (*Caporal* André Martin).

"'C— Wide-awake, energetic, bright little English-speaking Frenchman. (*Caporal* Pierre Cordier, whose English would come in very handy for me later).

"'L— Quite nice. A bit proud, handsome, nice enough, but not very sure of himself as a flyer, as yet. (*Lieutenant* André Laganne).

"'B— Amusing Algerian. Lazy, pleasure-loving. Avoids Boches if he can. Also trouble. (*Maréchal-des-Logis* Roger Busch).

"'P.P.S. Here's another I forgot, M—. Machine smashed now. Good flyer. Descended couple of Boches. Wounded in 1915 and was in hospital year and a half. Very swank, nice chap. (A few minor inaccuracies aside, I think I know this fellow…)'"

[Author's note: the above passages were quoted from *A Year for France: War Letters of Houston Woodward*, New Haven, Yale Publishing Assn., Inc., 1919, pp.143–146, with Martenot's comments in parentheses].

"We had several American volunteers in our *escadrille*," Martenot said, "but I didn't get to know them very well. Wallace Winter, Walter Shaffer and David Putnam were with us only a short time before they, Variot, Van Den Bosche and a few others were pulled out to help form a new *escadrille*, N.156, which was later equipped for a short time with the Morane-Saulnier AI parasol scout. I later learned of the death of my friend Variot in combat while in N.156, and of Putnam's exploits with them, and later at Spa.38 under Madon. I have read that Putnam is reported to have had a red Spad while with the 139th Aero Squadron, USAS. If that were so, he was probably influenced by Madon's practice.

"Austen Crehore and Houston Woodward stayed with us longer, but they stayed with Marinovitch most of the time. They were good friends, and besides, 'Marino' spoke English. I could not.

"There were groups of fellows who usually ate and flew together. I usually associated with Laganne, Pillet, Carbonnel, later de la Rochefordière and Bozon-Verduraz. Marinovitch,

Above: David Endicott Putnam was with N.94 only briefly before being transferred to the newly formed N.156. He went on to become one of the more prominent aces of the Lafayette Flying Corps. (Jon Guttman)

Above: Henry Houston Woodward's identification. The Lafayette Flying Corps volunteer would make his mark in N.94 and his letters home preserved many invaluable impressions of *escadrille* operations. (Jon Guttman)

Cordier, Van Den Bosche, Sainz and Ondet formed another sort of clique, later joined by Crehore and Woodward. Not that we had flights as other squadrons did. Pillet chose flight leaders arbitrarily, individually: He might say, 'Marinovitch, go up with Martenot and Cordier.' That was how our 'flights' were formed.

By the end of 1917, Martenot was firmly "sold" on the Spad VII. Inevitably a discussion of its merits led to commentary on its contemporaries—and especially the opposition.

"Of the enemy planes I encountered, the Pfalz D.III was difficult to maneuver, but we lost a lot of men to those planes. The Fokker triplane impressed us with its maneuverability and climb. You could not dogfight triplanes; the only way to fight them was with a hit-and-run attack through their formation, using the superior diving speed of the Spad to advantage. The Fokker D.VII was the best of the lot, with the easiest overall handling, but slower than the Spad. The Spad XIII was heavier, dove faster and by the end of the war, with engines as powerful as 300 hp, it was faster in level flight.

"German Rumpler two-seaters climbed much higher than we, and a lot more easily. It was hard to trap them. The LVG C.V was the best two-seater; it went high and fast for lone reconnaissances over us. We had to fly high altitude patrols to have the advantage over them, at the maximum altitude of the Spad – 6,500–7,000 meters. It was difficult to breathe at that altitude. And cold? Ha! Very cold! Our faces and legs were rubbed with petrol to aid circulation. We wore an insulated flying '*combinaison*' with rabbit fur lining, and a coat and gloves of goat fur. Despite the *combinaison*, you still got cold. My mustache would be covered with ice; even with the goggles, tears froze in your eyes."

Above: The pilot's identity is uncertain (Austen B. Crehore's grandson did not think it to be him), but this Nieuport 24bis provides a good view of N.94's original crab insignia as well as the style of individual numeral used in the late summer of 1917. (Rich Roberts)

Facing Page, Bottom: Nieuport 24, serial unknown, *Caporal* Austen B. Crehore, N.94, January 1918. This reconstruction is based on circumstantial evidence, suggested by Crehore's daughter, June Gulick: "My father had a 1940 or 1941 Packard sedan: Of course, we had it all through WWII, but after the war he held on to it, no matter how much we (family members) criticized it and begged him to replace it. In fact, he still had it when he passed away in 1962. This special automobile had custom license plates! He lived in Westfield, Union County, NJ until 1954, and the plates said 'UC 15.' From 1954 to 1962 he lived in Warren Township, Somerset County, NJ, and the plates said 'SCA 15' (my daughter still has that one; it is hanging on the wall in her garage). We have discussed the SCA and decided it might stand for "Somerset County Austen," in case 'SC 15' had already been assigned to someone else. In any event, because his automobile was so special and because he had special custom license plates for 20+ years, we think that '15' might have been the number of his plane. If no one else had that number, we think it is a good possibility." Combined with André Martenot de Cordoux's claim that he and most "old hands" in the *escadrille* kept the ID numbers they were originally assigned, this would seem extremely likely.

Nieuport 24, serial unknown, *Adjutant* André Martenot de Cordoux, July 1917. This is a reconstruction, based on Martenot's description, of the plane in which he scored his second official victory, over a German two-seater.

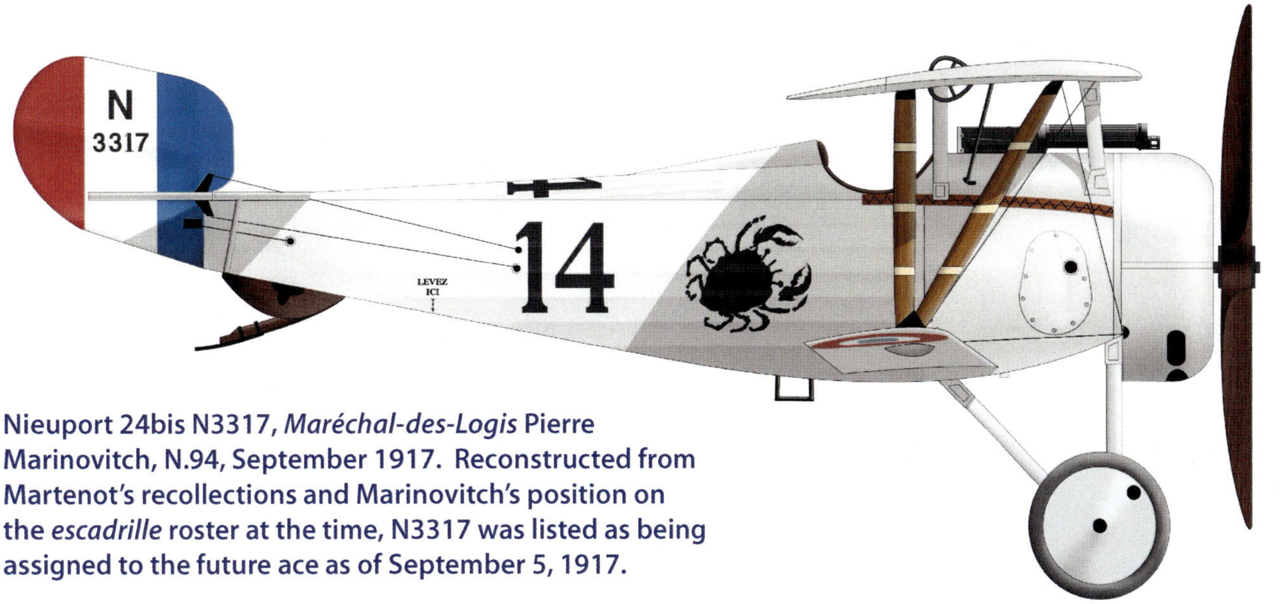

Nieuport 24bis N3317, *Maréchal-des-Logis* Pierre Marinovitch, N.94, September 1917. Reconstructed from Martenot's recollections and Marinovitch's position on the *escadrille* roster at the time, N3317 was listed as being assigned to the future ace as of September 5, 1917.

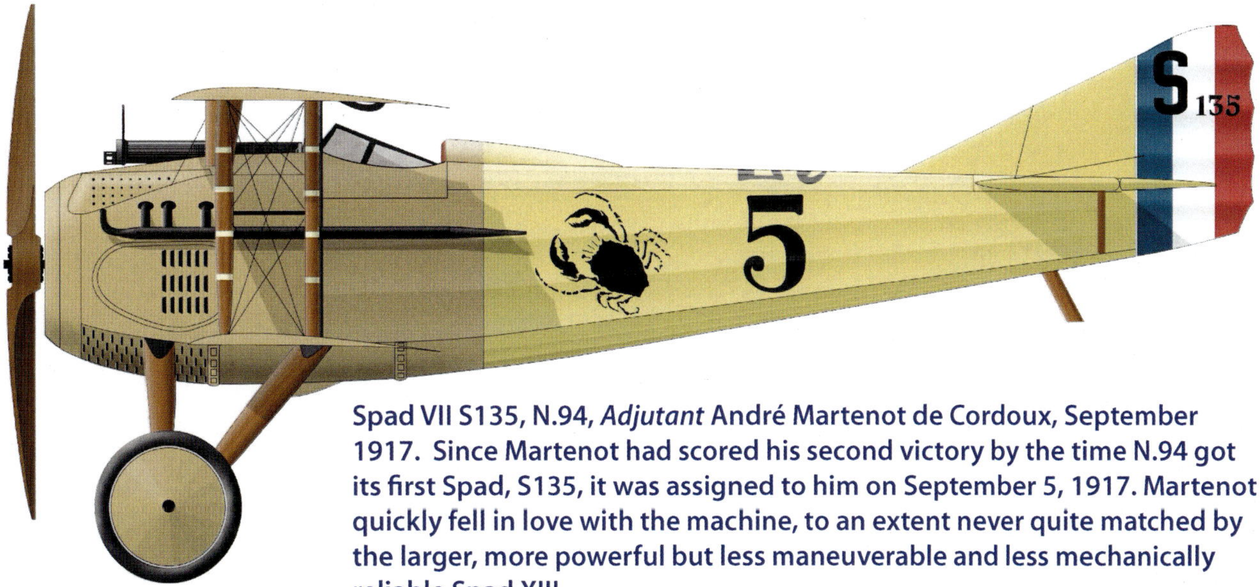

Spad VII S135, N.94, *Adjutant* André Martenot de Cordoux, September 1917. Since Martenot had scored his second victory by the time N.94 got its first Spad, S135, it was assigned to him on September 5, 1917. Martenot quickly fell in love with the machine, to an extent never quite matched by the larger, more powerful but less maneuverable and less mechanically reliable Spad XIII.

Nieuport 24, serial unknown, *Caporal* Austen B. Crehore, N.94, January 1918. This reconstruction is based on circumstantial evidence, suggested by Crehore's daughter, June Gulick; see facing page for explanation.

Chapter 2: Life at the Aerodrome

Left: Spa.94 takes advantage of good weather to improvise an outdoor mess hall in front of one of their Bessoneau hangars. Personnel include *Sergent-Méchanicien* Caritet (seated at near end of table), *Sergent-Chef du Bureau* Engonin (second from left), *Adjutant* Ondet (second from left on bench, facing camera) and *Adjutant* Marinovitch (at far end of table). (A. Martenot de Cordoux via Jon Guttman)

"Of course, not all our time was spent flying and fighting," Martenot continued. "Most of it was spent on the ground, living from day to day between our gambles with death.

"Of the ground personnel, two names return to me: *Sergent Méchanicien* Caritet, and our fat chief staff sergeant Engonin, who edited the history of 94 from day to day, of which the precise elements that I possess are especially useful for the present study. Engonin died of illness while in the *escadrille*, sometime late in 1918.

"Our food depended on supply and availability. Generally we had wine, cheese, vegetables and meat. Pilots ate better than the other soldiers.

"Between missions our fight was against boredom. We had our share of various forms of amusement: occasional ball games, cards, billiards. All the *escadrilles* had a bar!

"There was also the occasional leave to Paris, to get away from the Front and live it up for awhile. Personally, I almost never took leave to Paris – maybe once a year – until the Armistice.

"No account of a French *escadrille* would be complete without the obligatory mascot. Ours was a monkey. There was also a crow and a magpie. The CO and Marinovitch had dogs, and then there was my own dog, Fokker.

"I found Fokker abandoned in a canal near Reims and took care of him. He was a curious animal; he always came by and stayed with me when I was sick. He was always hanging around the mechanics all day, getting in everybody's way until they'd chase him off. But at night he'd always come into my tent and sleep by my bunk. He never lost the *escadrille* when it moved; he'd always stow away in a truck.

"One day, at Melette, we had an inspection by a general. Pilots and mechanics were lined up; visiting high-ranking officers went up and down the line, talking sharply to various officers while the general rode down the line on a horse.

"Suddenly, Fokker rushed out and bit the leg of the general's horse, which kicked and galloped off, the general hanging on for dear life, leaving everyone in a panic.

"When the CO came asking for me, I honestly told him that I didn't know anything about it. Fokker had disappeared, and while everyone was hunting for him for several days, they could not find a trace of him. Finally, they gave up and things returned to normal again. And who then turned up by my bunk, but Fokker!

"Other aspects popularly associated with World War I flyers are superstition and chivalry. Indeed, many of us had some superstition, such as three cigarettes on a match being bad luck. With flight itself still unreliable and survival in combat still very much a matter of chance, such thought was inevitable. But I personally indulged in no superstitions at all.

"In 1914 and 1915, with the warplane still a novelty, chivalry was common. Both sides exchanged information on the fate of pilots brought down on one side of the lines or the other. But as the war progressed, the airplane was transformed from a military toy to an effective weapon and aerial warfare became more earnest, and such courtesies became less and less common. By 1918, chivalry was dead...zero. On one occasion, our bombers, supported by fighter escorts, attacked the enemy. There ensued a furious and confused fight with 50 German aircraft, during which I saw one of our bombers forced down by several Boches – who then continued to machine-gun the crew on the ground. The knighthood of the air was short-lived."

Chapter 3: New Year Changes

On January 1, 1918, *Maréchal-des-Logis* Marinovitch got N.94's New Year off to an encouraging start when he sent a Rumpler down to crash north of Beine. On the same day, a new *escadrille* was formed – N.156, under *Capitaine* Emile Paumier – and a number of N.94's pilots were detached to form its cadre. The transfers included *Lieutenant* Alexandre Ottavy, *Sous-Lieutenant* Raoul Belloc, *Adjutants* Henri Variot and Pierre Barriot; *Sergents* Léon Van Den Bosche, Edouard Chatillon, Jean Garnier and Marcel Herbert; *Maréchal-des-Logis* Camille Grangeron; and *Caporaux* Jean Delaunay, David Putnam, Walter Shaffer and Wallace Winter.

"N.156, formed from N.94, was equipped briefly with the Morane-Saulnier AI in 1918," Martenot commented. "With a good motor, its maneuverability was superior, but it was difficult to pilot because it was hard to control the 150-hp Gnome monosoupape 9N rotary engine with which the combat model was equipped. Also, the wings were fragile, and it was replaced by the Spad in May 1918."

In addition to the mass exodus of transient personnel to form N.156, N.94 lost a seasoned veteran to another new *escadrille* in the making. On November 10, 1917 its XO, *Lieutenant* de Saint Sauveur, had finally moved on to take command of N.157, which officially formed 10 days later. In March 1918, when it became part of *Groupe de Combat* 21, all of whose aircraft were marked by a differently colored diagonal fuselage band, N.157 used a blue one, but its commander added to that a knight's sword entwined in mistletoe, part of the Saint-Sauveur heraldry. Re-equipping with Spads to become Spa.157 on May 5, the unit would finish the war with five enemy planes and a balloon to its credit, for the loss of five pilots.

While N.156 and N.157 prepared for combat, N.94's veterans broke in a new group of replacements while carrying on with the *escadrille's* regular duties. At this point the squadron's flying personnel were as follows: *Capitaine* Edouard Pillet, *Lieutenants* André Laganne and François de Lagerie; *Sous-Lieutenant* Georges Parizet, *Adjutant* André Martenot de Cordoux, *Sergents* Pierre Cordier, Auguste Cornille and Pierre Marinovitch, *Maréchaux-des-Logis* Edmond Bessières, Roger Busch and Joseph Ondet; *Caporaux* Eloi Bonnet, Austen Crehore, André Martin, Charles Nicholas and Houston Woodward; and *Brigadier* Miguel Sainz.

In the course of providing protection for a reconnaissance plane behind enemy lines on January 3, N.94 got into another engagement with the enemy, during which a bullet went through *Caporal* Charles Nicholas' arm and chest. Nicholas managed to reach French lines and force landed near Mourmelon. *Capitaine* Pillet and *Lieutenant* Parizet rushed to the field hospital, where the doctor expressed confidence that Nicholas would recover and the two officers returned to N.94 somewhat more reassured. At 2200 hours the *Médaille*

Above: *Caporal* Houston Woodward in front of his Nieuport 24, in which he scored a victory on January 6, 1918. (History of the Lafayette Flying Corps)

Militaire, just awarded Nicholas, arrived at the hospital, but the doctor held off informing his patient, wishing to see his commander award it to him more ceremoniously. A citation from the *Ordre de l'Armée* arrived at 0400 hours the next morning, but Nicholas' condition deteriorated in the course of the night, and at 0700 hours the doctor pinned the medal on him just before he took his last breath. He was buried at Mourmelon-le-Petite on January 5.

Pilots of N.94 wasted little time in avenging Nicholas' death. On January 4, Parizet engaged a German northwest of Somme-Py and claimed to have brought it down, although it was only credited as a probable. On the 6th, Parizet and Woodward attacked three Albatros scouts behind enemy lines and in a letter home on the 9th, Woodward described the details and excitement attending his first aerial victory:

"On January 6 three of us were to make an hour morning patrol. One of the French men had magneto trouble and couldn't go up, so Lieutenant Parizet and I started off to patrol the sector. There was nothing doing. Not a Boche in sight. So far as that goes, I haven't even seen a Boche our side for two weeks. We crossed the lines a bit at 4,000 meters and

Above: Lafayette Flying Corps volunteer Austen Ballard Crehore poses beside a Nieuport while training at Tours. (G. DeFreest Larner via Jon Guttman)

soon saw three Albatross monoplanes sailing along at about 3,300 meters. I didn't wait for Parizet, who was leading, to start for them, but piqued [dived] on one of them immediately. Parizet made a slight detour, then dove on one from the side, leaving the third, the leader, free. My Boche made a quick turn, so I redressed and began maneuvering to get behind and above him.

"Finally I got him where I wanted him, and piqued steep, shooting all the time. Parizet was then just ahead and above me, and I saw him shooting at a Boche who was maneuvering to attack me. He over-piqued eventually, and the Boche fired about twenty shots at me from the side and slightly below. He got so close I could see his face, and for a second I hesitated whether to turn on him or continue with the original one. He fell over on his side, though, so I let him go; I put my machine in a vertical nose dive, gaining tremendous speed, then redressed, and quickly overtook my fleeing Boche. Got within one hundred meters of him, and sent in a steady stream of bullets. When I was so close to him that I started to redress to avoid colliding with him, I saw him slowly slip over on the wing, then go into a slow *vrille*, and after a few maneuvers to keep him always under fire, I saw he had been hit, and made a vertical spiral to watch him vrille down to the ground. It was now at about 2,500 meters, and the other two Boches about 1,000 meters below. Parizet had remained at 3,000 meters, but I decided to take my chances with the other two,

Above: "Les Tangos." Albatros D.Vs of *Jasta* 19, showing their yellow and black-banded tailplanes. D.V 2111/17 in the foreground, bearing the initial "M" on the fuselage side was flown by *Vizefeldwebel* Martin Mallmann, whose death north of Beine on January 19, 1918 was jointly credited to Crehore and Marinovitch. This was not the last time that *Escadrille* 94 would encounter *Jasta* 19. (Arthur Rahn album/National Museum of the US Air Force)

so threw my machine on her side and dropped seven hundred meters like a plummet in a couple of seconds. Both Albatross immediately continued their piquing, I followed one as low as 1,000 meters but dared go no further after him. Then the fireworks began. One thousand meters is extremely low for five kilometers inside the German lines, and the air became black around me with their anti-aircrafts. I couldn't go in a straight line and, as there was a heavy head wind, it took me ages to get inside our lines again.

"That is all the description I can give of my first fight. It was very thrilling, and the most wonderful sport I have ever participated in. I was in danger only the time when the Boche fired at me, and then somehow it seemed so funny I burst out laughing. I had always rather dreaded my first combat, but there's nothing nervous or rattling about it. It was more like practice at target shooting than anything else. The aim has be be carefully timed and corrected. There is a tremendously exhilarating thrill about it, however, and the passion of the hunt."[3]

Woodward took part in a larger patrol that afternoon, but became bored and dropped out of formation, venturing 10 kilometers into German territory seeking new game. At about 3,600 meters he found a lone fighter coming his way, but identified it as a fellow Nieuport and fell in with it. For three minutes the two planes cooperated, the pilots scanning the sky to each side with Woodward 150 meters behind and 50 meters above his newfound wingman, before he happened to notice crosses on the upper wings. Having already pulled what he described as "one of the dumbest, biggest bonehead stunts in the war's aviation history,"[4] Woodward added another error by carrying out a hasty direct diving attack, rather than slipping

The Tangos

Asked about the *Jagdstaffel* locally known and often mentioned in various French and American pilots' memoirs as the "Tangos," André Martenot de Cordoux commented: "I didn't keep a particular memory of the German squadrons, because certain ones came into our region generally in periods of major land attacks by Germans or French. The 'Tangos' seem to have stayed longer than the 'Reds' and 'Checkerboards,' as well as the 'Greens.' Their identities were unknown to me. I never had personal contact with my adversaries. On the rare occasions when I would have been able to meet them, when they had been brought down wounded and taken prisoner by ground troops, I would learn that they had already been evacuated to hospitals in the rear."

Caporal Austen B. Crehore's first victory on January 19, 1918 coincided with the death of *Vizefeldwebel* Martin Mallmann of *Jasta* 19 over Vaudetrée, a short distance north of Beine. At that time, *Jasta* 19 was equipped with Albatros D.Vs whose varnished plywood fuselages may have looked like the orange "tango" color described by so many of their French adversaries, in addition to which the tails were painted in chrome yellow with black bands. Circumstantial though that evidence may be, it makes a convincing case for *Jasta* 19 to have been the "Tangos" reported by so many French adversaries in its sector.

Born in Boppard on July 28, 1892, Mallmann began his combat flying in two-seaters with *Flieger Abteilung (Artillerie)* 226 before retraining on fighters and being assigned to *Jasta* 19 on July 6, 1917. He opened his account less than two months later by burning a balloon at Pontavert on September 3, after which he shot down a Caudron over Villers on September 19, and another Caudron over Berry-au-Bac on December 7. On December 15 he claimed an AR.1 near Craonne, but it was not confirmed. Then came that fateful dogfight on January 19, 1918...

Jasta 19 operated in the Reims sector throughout most of 1917, and would reappear there in 1918 in affiliation with *Jagdgeschwader* II, at that time operating Fokker Dr.I triplanes with the same yellow and black tailplanes, before eventually re-equipping with even more fearsome Fokker D.VIIs.

Above: On the same day Crehore and Marinovitch downed *Vizefeldwebel* Martin Mallmann, former N.94 member *Adjutant* Henri Variot, now with N.156, was killed in action. His death was as much a blow to his former comrades-in-arms at N.94 as it was to his squadron mates at 156. (*La Guerre Aérienne Illustrée No.*73)

under the enemy plane's tail and attacking from below. The Albatros D.Va went into a dive and Woodward last saw it at 1,000 meters, unwilling to turn and engage him. For his own part, he returned home through another barrage of anti-aircraft artillery fire, which he easily evaded by judicious zig-zagging.

Back at Melette, Woodward learned that the Albatros he had fought that morning had been confirmed as coming down near Manre. *Capitaine* Pillet congratulated him on his first victory, but also told him that in operating so far behind German lines with less than three aircraft he and Parizet had disobeyed a standing order and forbade them from doing so again.

On January 12 one of N.94's new replacements, *Caporal* Auguste Cornille, was coming in from a training flight when his plane suddenly stalled and nosed into the field. He had only been with the *escadrille* four days, but he emerged without injury from a wreck that had to be taken away for major repairs. On the 15th *Capitaine* Pillet left for Aircraft Park No.4 to take a four-day course on the Hispano-Suiza engine—an exciting indication, from the standpoint of N.94's pilots, that they would soon be replacing their aging, fragile Nieuports with fast, sturdy Spads.

Although Woodward had been the first American to make a material contribution to N.94's record with his January 6 success, he was not to be the last that month. Among Martenot's records was the item: "19 January, 1918 – *Maréchal des Logis* Marinovitch and *Caporal* Crehore shot down a 'Tango' Albatros in their lines, north of Beine."

Pierre Marinovitch had rather more to say of that fight. "I got to know Crehore and Winter awfully well," he said in an interview, "and we were the three best friends in the world. Crehore wanted to get boches terribly hard; the only trouble was we had Nieuports while nearly every other squadron had Spads, and we were terribly handicapped. Crehore and I were flying every day, never could we find a Boche. December passed, and January, and poor old Crehore was nearly wild. By this time Putnam had got his first, and on the 15th [*sic*] of January Crehore and I decided we must see one, so we went far into the enemy's lines and waited. Finally I spotted a Boche alone, and a big patrol, but much farther away. I put full motor on, outdistancing Crehore (he had a cylinder off his motor, which of course I did not know). I got into a fight with the Boche, but he, having a much superior machine – it was an Albatross – all painted 'tango,' soon got the better of me, and was just going to bring me down, when Crehore by clever maneuvering caught him and the Boche fell vertically, losing a bit of his plane. We were twenty kilometers in Germany, and we still had to get back before the big patrol got us. No matter, Crehore followed him to the ground! This was the first Boche Crehore had ever seen, also his first official victory, and I'll never forget how he saved my life with a rotten machine and a motor on the bum!"[5]

Crehore's first victory—officially shared with Marinovitch as his fifth – coincided with the death of *Vizefeldwebel* Martin Mallmann of *Jasta* 19 over Vaudetrée, a short distance north of Beine. It did not go unnoticed at higher levels. Ten days later, Crehore was awarded the *Croix de Guerre* with the following remarks from the commander of the *IVème Armée*, *Général* Henri Gouraud: "Young fighter pilot, fine and bold, has attacked his adversaries since his first sorties, downed one far in their lines on 19 January 1918."[6]

Crehore was not the only pilot to make news on January 19. N.156, flying Nieuports while awaiting its Morane-Saulnier parasols, logged its first victory, but also suffered its first loss. Woodward, who had seen several friends off to the new unit, wrote of the day's activities in two letters home.

"Today was the first good day since the 13th. Everyone was terribly fed up with the loafing, so very restless. The result was that it has been a very memorable day. Everyone in the group flew, and judging from all the Boche machines I saw, I guess all the Huns were up too. This morning Winter and Putnam flew with a French Lieutenant. They penetrated the German lines six or seven kilometers, then things happened so fast no one can give a clear account. Winter was piqued on by two Boches, probably Fokkers, and threw himself around like a cork on rough water. How he got away he doesn't know, but he did every acrobatic known, and somehow escaped. Belloc attacked a Fokker, then found himself attacked by two, and by jumping around got away. Putnam got separated from the others, attacked two Fokkers and brought one of them down, the Boche plunging into a forest. Unfortunately it was not hanologated [*sic*]."[7]

On January 21 Woodward noted: "Put's Boche was hanologated. Some infantry observed the chute."[8] Woodward's term for confirmation was actually based on the French "homologué," and "chute" in this context was the French word for "fall." The "Fokker," which fell near Nogent l'Abbesse was credited to *Sous-Lieutneant* Belloc and *Caporal* Winter as well as Putnam. Their victim was probably *Unteroffizier* Richard Kade of *Jasta* 50, who crashed to his death at la Neuville, about 20 kilometers northeast of Nogent l'Abbesse.

Woodward had other events to record in the same letter. While attacking a two-seater he became separated from his flight and was chased back toward the lines by three enemy fighters. He then encountered one of his wingmen, *MdL* Edouard Bessières, spotted another enemy two-seater at about 1500 hours and dived on it, unaware that Bessières had not seen it. In consequence, he suddenly found himself alone against his quarry and two other German planes that joined it, and beat a discrete retreat through a hail of anti-aircraft shells. Meanwhile, the French had received reports of a Nieuport going down in flames at 1515 hours and assumed it to have been Woodward, especially since Nieuports normally could stay up for two and a quarter hours, and Woodward stated that he finally returned after two hours forty minutes "with not enough essense to wet the tank." This time, he wrote, "The captain was absolutely pale, and gave me hell for ten minutes straight."[9]

The pilot of the Nieuport reported lost that afternoon was subsequently identified, and the news struck painfully close to home in both N.94 and N.156. That included Woodward, as he concluded his January 21 letter:

"Poor old Variot. He used to be in 94, but transferred to 156 when that escadrille was formed. He still lived with us, however, and everyone liked him. Very good pilot, old flier, and had two Boches to his credit. Loin chez eux [far in their territory] he became separated from his two companions, and no one has ever seen him since. It was phoned in that a Nieuport was seen fighting six Albatross for five minutes, and then descended in flames. It was impossible to do otherwise. He should not have tried to fight. If I ever get in a tight place chez eux I am going to pretend that I am hit, and drop like lead. When very near the ground I shall redress, and just skin the soil full speed, leaping trees and fences, and thus with comparative safety reach chez nous. I have practiced it, and can get away with it I am sure. This will be a great safeguard, and a pretty sure preventative from joining the daisy pushers. That same day Marinovitch and Crehore together brought down one of the Tangos – Germany's famous circus escadrille. This makes Marinovitch ace, and his third machine within a month."[10]

Reportedly killed near Moronvillers at 1610 hours, Henri Variot had been awarded the *Médaille Militaire* and the *Croix de Guerre* with two palms and one gold star before his last fight. He was probably the victim of *Leutnant* Fritz Pütter of *Jasta* 9, who claimed a Nieuport west of Tahure, about 15 kilometers southeast of Moronvillers, for his ninth victory.

Above: Variot was probably the ninth victory credited to *Leutnant* Fritz Pütter of *Jasta* 9. Later commanding *Jasta* 68, Pütter brought his total to 25 before being severely burned when his own tracer ammunition caught fire, and died of his injuries on August 10, 1918. (Aaron Weaver)

On January 22 Martenot led Marinovitch, Woodward and Crehore in a strafing attack on German trenches near Tahure. Three days later, N.94's log noted that a patrol composed of *Lieutenant* Parizet, *Lieutenant* Laganne, *MdL* Bessières and *Brigadier* Bonnet were attacked by an enemy single-seater, which they shot down near Fontaine-en-Dormois, Credit was shared by all four pilots.[11]

The day's success was particularly encouraging for Laganne, who had been with the unit little more than a month. Born on June 2, 1894, he was a student when war broke out, and on September 2, 1914 he was mobilized into the 11ème *Régiment d'Infanterie*. Rising to *lieutenant*, he decided to enter aviation on July 25, 1917, and earned his brevet at Avord on September 3. After further training at Pau and Cazaux and GDE, he was earmarked for N.156 on December 15, but was sent to N.94 instead. With one victory, albeit a shared one, to his credit, Laganne was undoubtedly feeling more confident than he had seemed when Houston Woodward first met him.[12]

Above: *Caporal* Walter John Shaffer poses in the cockpit of his Morane-Saulnier AI parasol fighter, MS1591, which bears his personal number 11 but has yet to have MS.156's insignia applied to the fuselage side. Much was expected of the Morane, but its combat career was brief and disappointing. (W.J. Shaffer Album)

On January 26, *Brigadier* Sainz was forced to land two kilometers north of Somme-Bionne, overturning his plane but surviving the crash. January 27 brought another aerial victory for an N.94 patrol, but one that was only counted as a "probable." Sister unit N.156 did better, dispatching an Albatros two-seater in flames over Vaudesincourt, credit being shared by *Capitaine* Paumier, *Lieutenant* Belloc, *Adjutant* Barriot and *Caporal* Putnam. Once again, however, there was a price to be paid for the new *escadrille*'s second confirmed victory, as the reconnaissance plane's fighter escorts drove down Belloc in Nieuport 24 N5964 near Dontrien, about five kilometers south of Vaudesincourt, at 1713 hours German time, credit again going to *Leutnant* Pütter of *Jasta* 9, for his 10th victory.

Lieutenant Raoul Louis Amadée Belloc was born in Le Havre on January 6, 1892, and began his military service with the *9ème Régiment des Cuirassiers* on October 1, 1912. He subsequently served in the 2nd Colonial Artillery and 17th Infantry regiments. Commissioned a *sous-lieutenant* on September 8, 1915, he transferred to aviation on March 21, 1917, and qualified for Military Pilot Brevet No. 6284 on May 9. Promoted to *lieutenant* on September 8, he joined N.94 on December 1, but only in preparation for the formation of N.156. He received the *Croix de Guerre* with one gold star, but soon after force landing his stricken plane, Belloc died of his wounds.

On January 29, *Général* Gouraud arrived to pin the *Croix de Guerre* on *Capitaine* Pillet, who had been cited in the *Ordre de l'Armée* for his past reconnaissance work and his more recent leadership of his new fighter command, and on *MdL* Ondet. He also awarded *Sergent* Marinovitch the *Médaille Militaire*.

The next day orders arrived for N.94 to depart Melette for to the aerodrome at Villeneuve les Vertus, which it did on the 31st. There it was incorporated into *Groupe de Combat* 18 under *Capitaine* Jacques Sabattier de Vignolle, along with *Escadrilles* Spa.48, N.153 and N.155. On February 4, GC.18 was combined with sister groups GC.15 (*Escadrilles* Spa.37, Spa.81, Spa.97 and R.46) and GC.19 (*Escadrilles* Spa.73, Spa.85, Spa.95 and Spa.96) to form an even larger formation, *Escadre de Combat* 1, under the overall command of *Chef de Bataillon* Victor Menard. N.94's war was about to begin in earnest, and its equipment was soon to reflect that fact.

February 2 had brought a significant change to *Adjutant* Martenot's status, when he was commissioned as a *sous-lieutenant*. The next day saw the long-anticipated order come down for 94 to send its Nieuport 24s and 27 to Melette and its Nieuport 24bis to GDE. After training on Spad VIIs at GDE, the pilots were to pick up their own new Spads at Dugny, and then fly them back to Villeneuve.

One squadron member would have to wait before taking his

new mount into action, however, when Crehore was diagnosed with appendicitis on February 7 and after spending time in the Hôpital de Notre-Dame d'Epernay, was shipped home on March 11 for two months' convalescence in the United States. As things would turn out, though, Spa.94 had not seen the last of Austen Crehore.

N.156 also began replacing its Nieuports at that time, as it received its first Morane-Saulnier AI from Plessis-Belleville on February 4. Bearing the military designation of MoS.27C.1, the new parasol fighter already had a sterling reputation. Among the first pilots sent to ferry one in was Dave Putnam. Walter Shaffer, who came to pick his plane up on the 8th, met Putnam and wrote: "He was greatly elated over the speed and climbing ability of the Morane and told me how nicely it worked in the air." Shaffer's impression was that "it was tiny, but it is so beautifully proportioned and the lines so racy-looking that a more beautiful 'zang' was never built."[13]

N.94's spinoff sister *escadrille* was redesignated MS.156 on February 9, but its pilots' elation at their new mount gave way to trepidation on 26 February, when their executive officer, *Lieutenant* Jean Toutary, lost his wings while performing aerobatics over their aerodrome near Châlons-sur-Marne. On March 8, five of MS.156's planes encountered two German two-seaters over the lines and two of them dove to attack. One of the Morane-Saulnier pilots discovered that his gun was jammed and pulled up to clear it. The other, *Caporal* Winter, continued his dive until his wings suddenly folded up and he plunged into the ground at Pont-Faverger.

Wally Winter's death saddened the *escadrille* and raised a lot of uneasy questions regarding the MoS.27's structural integrity. Postwar German records reveal a curious anomaly – a Morane-Saulnier was credited to *Leutnant* Julius Keller of Royal Saxon *Jasta* 21. Had he come to the rescue of the two-seaters and fired on Winter from above and behind, unnoticed by Winter's French squadron mate? Whatever the actual cause, the high expectations that attended the parasol fighter's delivery plummeted as fast as Winter's broken plane.

Most of MS.156's pilots resumed flying missions in Nieuport 27s, but at least one, the newly promoted *Sergent* Putnam, was not so easily intimidated. On March 14, he attacked and claimed an Albatros near Nauroy, although it was only counted as a "probable." On the following day he downed a Rumpler over Beine, which was confirmed as his – and MS.156's – third victory. As combat around Reims intensified, Putnam grew more aggressive, but his tendency to fight deep in enemy territory prevented most of his claims from having enough air and ground witnesses to be confirmed in accordance with French regulaions. For example, he claimed two enemy fighters on April 12, a triple victory on the 23rd and another on May 15, but all went in his record as "probables."

Meanwhile, further crashes had convinced the French that the Morane-Saulnier AI was not suitable for front-line service. On May 20, Shaffer wrote: "And now we have changed airplanes again, getting Spads this time… The Moranes, I am sorry to say, have been given up, owing to their weak construction,

Above: Like David Putnam and Walter Shaffer, Lafayette Flying Corps volunteer *Caporal* Wallace C. Winter briefly served in N.94 before helping establish N.156, but his combat career proved tragically brief. (*History of the Lafayette Flying Corps*).

which could not stand the strain '*chasse*' work entails. I say sorry, because not only was it fast, but so small that as one pilot said, it would be maneuvered around a clothes pin…"[14]

On June 1, Putnam and Shaffer transferred to Spa.38, where, under Georges Madon's leadership, both pilots came into their stride. Shaffer was credited with a balloon on August 4 and a fighter on August 21, before being brought down and taken prisoner on October 10. By the time Putnam transferred to the U.S. Army Air Service on June 24, he had raised his official score to nine and had received a first lieutenant's commission on June 8. Upon being assigned to the 139th Aero Squadron, he was made a flight leader.

Spa.156, as it became when it acquired Spad VIIs and

Morane-Saulnier's Parasol Fighter

Above: Morane-Saulnier AI parasol fighter, apparently a single-gun MoS.27C.1. The AI offered sparkling performance and maneuverability, but two fatal wing failures led to its replacement by the sturdy Spad.

When N.94 was split to form N.156 at the start of 1918, the new *escadrille* was earmarked to receive a new fighter: the Morane-Saulnier AI. Developed in the summer of 1917, this latest in a long line of monoplanes was of conventional wood and fabric construction, with wire stringers fairing the fuselage into a circular shape from the metal cowling to a point at the tail. The slightly swept-back upper wing was mounted above the fuselage in a parasol arrangement by means of a truss-like strut arrangement that yielded a safety factor of 8.5 during static testing. Power came from a 150-horsepower Gnome monosoupape 9Nb engine and during flight tests at Villacoublay in August 1917, the prototype climbed to 3,000 meters (9,840 feet) in 7 minutes and 45 seconds, and reached a speed at that altitude of 134.5 miles per hour. That performance was considered phenomenal for a rotary-engine fighter, and since the Spad XIII was experiencing trouble with its geared Hispano-Suiza engine, the French ordered the Morane-Saulnier into production, with anywhere from 1,100 to 1,300 being built. The earliest AI variant, armed with a single synchronized Vickers .303-inch machine gun, was designated the MoS.27C1, while a later twin-gun version was called the MoS.29C1.

The Morane-Saulnier's reputation preceded it when N.156's pilots learned that they would be the first to receive the new fighter on February 4, 1918. By February 9 it was fully reequipped and redesignated MS.156. Two other *escadrilles*, MS.158 and MS.161, were fully equipped with MoS.27s by March 4 and February 21, respectively. One of MS.158's Americans, *Caporal* Rufus R. Rand Jr., was equally impressed after a few frontline flights, stating that with a few alterations, it could be one of the best fighters on the Western Front.

Jubilation soon gave way to trepidation, however. First *Lt.* Jean Toutary, MS.156's executive officer, lost his wings and his life while stunting over the aerodrome on February 26. Then *Caporal* Wallace Winter's wings folded up during a combat on March 8. With the airframe's ability to stand up to combat stress in question, most of MS.156's pilots resumed flying missions in Nieuport 27s, with the notable exception of *Sergent* David Putnam, who used one to down a Rumpler for his – and MS.156's – third victory on March 15. He and *Caporal* David Guy were still claiming aerial victories in the parasol fighters, albeit all too far in German lines to confirmed, as late as May 15. On that same day, however, *Caporal* Émile Boucheron of MS.158 crashed while trying to land his Morane-Saulnier at Maisonneuve. Aside from its suspect wing structure, the plane suffered from the vagaries of its Gnome monosoupape engine, which would also bedevil the Nieuport 28C1 during its brief combat career with the U.S. Army Air Service. Morane-Saulnier installed a 170-hp Rhône 9R in an MoS.27 in place of the Gnome, but it failed to rekindle interest on the part of the *Aéronautique Militaire*.

The first twin-gun MoS.29s arrived at MS.156 on May 18, but by then their cause was lost. On May 19 *Escadrilles* 158

Above: Morane-Saulnier AI parasol fighter, apparently unarmed.
Below: Morane-Saulnier AI twin-gun MoS.29C.1 parasol fighter.

and 161 replaced the last of their MoS.27s with Spad VIIs, and MS.156 did so the next day.

Subsequent production Morane-Saulnier AIs had their wing supports bolstered with extra lift cables, but for the rest of the war they were relegated to the advanced training ("*Entrainement*") role under the designation MoS.30E1. Armed with one Vickers gun, the MoS.30 also had a 120-hp Rhône 9Jb or 135-hp Rhône 9Jby engine. A more purpose-built variant, the MoS.31bisE1, had a downrated 90-hp Rhône and a reduced fuel capacity. As a trainer the AI saw extensive use in the French fighter schools and 51 were purchased by the USAS for its training facility at Issoudun. Test pilot Temple N. Joyce showed what the plane could do when he put one through 300 consecutive loops, but his report also reiterated the Gnome monosoupape engine's propensity for catching fire if not handled carefully.

The Morane-Saulnier Mos.30E1 saw limited postwar use in other countries. Belgium acquired three, one of which bore the thistle insignia of its top fighter squadron, the *9ème Escadrille*. Czechoslovakia and Switzerland each bought one, and one operated at the French flying school in Warsaw, Poland, in 1920. Japan evaluated one for possible use as a trainer in 1922, but ordered no more.

Above: This single-gun MoS.27C.1 parasol fighter was one of 15 assigned to MS.156 in February 1918.

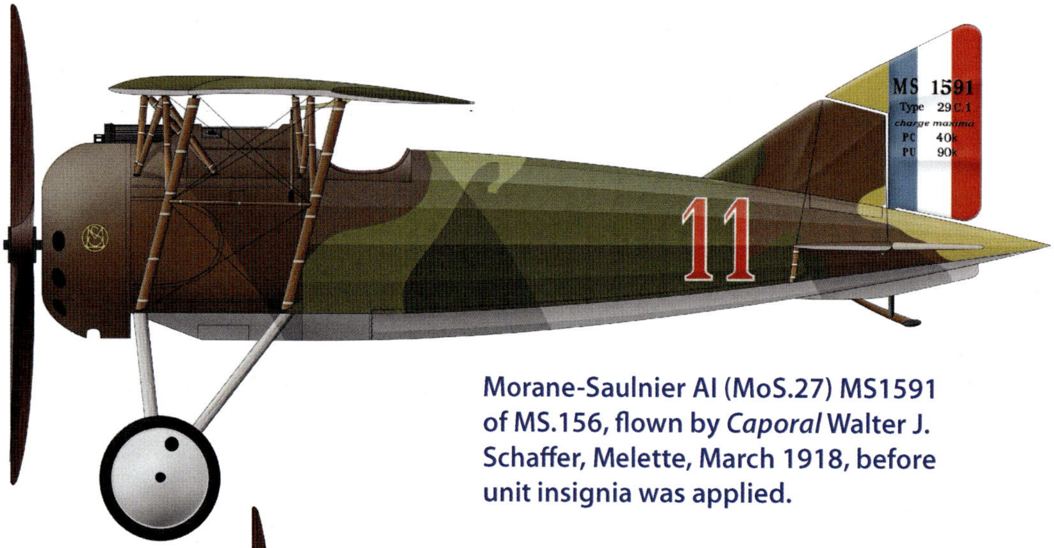

Morane-Saulnier AI (MoS.27) MS1591 of MS.156, flown by *Caporal* Walter J. Schaffer, Melette, March 1918, before unit insignia was applied.

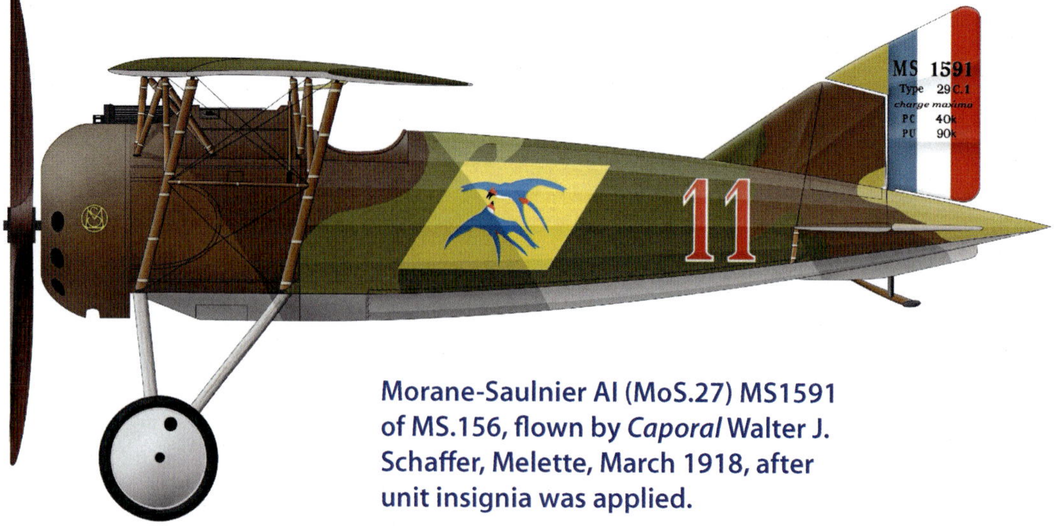

Morane-Saulnier AI (MoS.27) MS1591 of MS.156, flown by *Caporal* Walter J. Schaffer, Melette, March 1918, after unit insignia was applied.

Above & Below: Unarmed Morane-Saulnier AI MoS.30E.1 dedicated trainer version with additional bracing wires.

Above: *Adjutant* Léon Van den Bosche beside his de Marçay-built Spad VII, S7127, displaying the two blue martens of *escadrille* 156. One of N.94's original members, Van den Bosche help found N.156, flying Nieuports, Morane-Saulniers and finally Spads. (SHAA B87.1485 via Jon Guttman)

Above: Painting of *MdL* Roger Busch, who was shot down and made POW on 12 April 1918, probably the ninth victory for *Leutnant* Ulrich Neckel of *Jasta* 12. (Patrick Vandenbussche via Albin Denis)

XIIIs, was never quite the same without Putnam. Although it continued to claim enemy aircraft for the rest of the war, only three were confirmed – and Putnam had figured in all three of them.

On March 3, 1918 N.94 was redesignated Spa.94, officially signifying its transition to Spad VII and XIII fighters. March 4 saw de Lagerie promoted to a temporary rank of full lieutenant, as of February 26. On the 7th, Marinovitch learned that he had been promoted to *adjutant* as of February 20. Moreover, with five victories to his credit, the 19-year-old paladin of N.94 found himself being referred to as the "*Benjamin des As*" by the French press, in reference to his youth. Cornille was promoted to *sergent* on the 13th. On the 22nd Martenot received a rather belated *Croix de Guerre Belge* for valor and for wounds suffered when his infantry unit operated alongside the Belgian Army during the 1914 Battle of the Yser.

One month later Marinovitch, Martenot and the pilots of Spa.94 would face their greatest challenge, as the Germans launched their final great offensive of the war.

Chapter 4: Death Duel Over the Champagne

Above: *Adjutant* Martenot's Janoir-built Spad VII S3190 is in the foreground of this lineup of Spa.94 planes with the combination crab and *La République escadrille* insignia. This was probably the plane in which he shot down an Albatros D.V for his third victory on April 1, 1918. (Greg VanWyngarden)

At 0440 on the morning of March 21, 1918, the Germans launched what they called the *Kaiserschlacht*, a general offensive that represented their final bid to take Paris and crush the Allied armies on the Western Front before the American Expeditionary Forces, or AEF, could arrive in sufficient force to shift the fortunes of war irretrievably in the Allies' favor. The opening phase, codenamed Operation Michael, began with a concentrated bombardment by 6,000 German guns against the British Third and Fifth armies along the 42-mile-front south of Arras. The British positions were then assaulted by the *2. Armee* under General Georg von der Marwitz, the *17. Armee* under General Karl von Bülow, and the *18. Armee* under General Oskar von Hutier. Smashing their way through the center of Lieutenant-General Sir Hubert Gough's Fifth Army, the storm troopers of Hutier's *18. Armee* reached open country in less than 10 hours. The mastermind behind the German push, General Erich Ludendorff, ordered Bülow's *17. Armee* to concentrate on seizing the critical rail center of Amiens, while Hutier was to direct his *18. Armee* along the right flank of the reeling British Fifth Army, in order to drive a wedge between it and French *Général* Denis Duchêne's VIème *Armée* to the south. It was only the first of a succession of hammer blows that the Germans would land to the Allied armies over the next several months.

Initially, GC.18 was assigned to the defense of Paris, which had come under attack by enemy bombers. Then the German armies shifted their drive southward to advance 45 kilometers, seizing Albert and Montidider in the process, and GC.18 was among the *groupes* redeployed there to oppose them. In so doing the French encountered some of the best squadrons in the German air service, veterans from the British sectors at Flanders and Amiens. One antagonist that Spa.94 would encounter with particular frequency was *Jagdgeschwader* II, which like GC.18 was a permanent grouping of four squadrons to be concentrated wherever local air superiority should be needed. Formed in February 1918, JG.II was comprised of *Jastas* 12, 13, 15 and 19.

On March 26, *Escadre de Combat* 1 arrived at Plessis-Belleville. On that same day the combat report stated:

"*Adjutant* Marinovitch, part of a patrol with *Lieutenant* Laganne and *Caporal* Martin attacked an enemy biplane around 1130 hours just touside Caurel. The combat began at about 3,000 meters and ended at 400 meters. The enemy biplane fell in a spin, damaged with black smoke. In Martinovitch's machine there was a bullet through the stabilizer and one through the rudder. The enemy biplane, probably shot down, fell into the middle of a thick mist.

"We have learned," the report added, "that *Adjutant* Garaud

Above: A painting of *Caporal* Houston Woodward, by Princess Marie Eristoff. Woodward was killed on April 1, 1918. (Jon Guttman)

of Spa.38 was grievously wounded in a fight with an enemy plane that he shot down. He received a bullet in the lung. It is hoped that he nevertheless will recover."[15]

At first credited as a probable, the crash of the two-seater claimed by the Spa.94 trio was later confirmed and credited to all three participants after the war, giving Marinovitch a belated 22nd victory and Laganne a never-formally acknowledged fifth.[16] *Adjutant* Hector Garaud's 12th confirmed victim, on the other hand, came down in French lines near Châlons-sur-Marne. Rumpler C.VII 6484/17 of *Fl.Abt.(A)* 270 was recovered and its crew taken prisoner. Another of Georges Madon's disciples at Spa.38, the 20-year-old Garaud would indeed recover from his wound to shoot down a Fokker D.VII for his 13th and final victory on August 13.

The next day, GC.18 began flying daily missions in support of the Allied forces desperately fighting to stem the German offensive. On March 30, 18-year-old *Brigadier* Pierre Chan joined Spa.94. Hailing from Paris, his original surname was Cahn, but he stated in a 1979 interview that his family had chosen to alter the letter order to make it less obviously Jewish-sounding during the wave of anti-Semitism that had swept over a disturbing percentage of France's populace in the course of the trial and conviction for espionage, imprisonment and ultimate exoneration of *Capitaine* Alfred Dreyfus, between 1894 and 1906.

April 1 saw a high level of activity, distinguished by an equal measure of triumph and tragedy for GC.18. Martenot remembered it for the prominent role he played in it, as well as the loss of a well-liked squadron mate:

"In the morning a patrol of six planes had to escort some bombers: *Caporal* Woodward attacked a *Drachen* (observation balloon) without result. On one of my lone sorties in my Spad VII, I ambushed a *Staffel* of Albatros over Chaulnes in a classic example of diving Spad attack, shot one of them down and dove away. It was later indicated in the report that I had shot down an Albatros with the participation of *Lieutenant* Béchon, a pilot of Spa.155. *Lieutenant* Béchon was not on patrol with me; being in the vicinity of the combat, he had put his nose in our direction and fired from afar. His presence was not withheld for the homologation, which was granted to both of us."

Even in 1979, Martenot did not consider *Lieutenant* René Béchon's participation convincing enough to warrant his share in that particular kill. "I deplored the manner in which planes downed by our patrols were too often credited to everyone," he remarked. "The German method (one pilot credited per victory) was more equitable, but nevertheless not perfect. It is difficult to precisely trace the origin of the effective bullets in some combats. I believe that our method was taken to encourage and stimulate those who didn't obtain any success, though nonetheless taking part in the risks by their presence in the fight, which gave cause for regrettable abuses in certain *escadrilles*."

Martenot's victim was probably *Leutnant* Richard Grüter of *Jasta* 17, who the Germans recorded as shot down and killed over Montdidier, near Chaulnes. *Lieutenant* Béchon and his Spad VII were posted missing in action later that same day, possibly the Spad credited to *Leutnant* Wilhelm Buchstett of Royal Bavarian *Jasta* 79. Under the circumstances, it is likely that members of Spa.155 petitioned their higher-ups for their late comrade in arms to share credit in the earlier victory.

Béchon was not the only loss that GC.18 would suffer that day. "In the afternoon of April 1 we had several more fights," Martenot concluded. "Woodward, part of a six-man patrol, had several combats, and did not return, missing over Montdidier."

Not for the first time, *Capitaine* Pillet wrote a letter to the family of a lost *escadrille* member on April 9. This time it was to Woodward's father:

"I duly received your favor of the 5th inst. and beg to send you at once the information I have on *Caporal Pilote* Houston Woodward who has disappeared and is much regretted by my *escadrille*.

"The first of April, in the afternoon Woodward was sent on a patrol in the enemy's lines; he was seen several times during that patrol by myself. The French patrol having attacked a certain number of enemy airplanes, there was a fight after which *Caporal* Woodward disappeared.

"The fight took place in the North and North-West region of Montdidier at a low altitude (about 1,000 meters) and two hypotheses seem possible: *Caporal* Woodward was brought down during the fight, wounded or killed, or else an accident occurred to his machine obliging him to land in the enemy's lines. The thick fog and the clouds which existed on that day prevented us from learning more of him.

"*Caporal* Woodward was in a Spad type machine, bearing No. 1419 with motor No. 10.828.

"Thanking you in advance for any inquiries you may wish to make on his account, we beg to remain,

"Yours very sincerely,

"Signed by the Commandant of *Escadrille* Spad 94."[17]

Woodward had gone missing during an evening patrol, as did *Sergent* Cornille. The next day Spa.94 got a telephone call from Cornille, who reported that he had force landed at Ormoy-le-Davien due to carburetor trouble and was in need of fuel to get back. Once mechanics had put his engine back in order, he rejoined the *escadrille* at noon. Woodward remained unaccounted for.

On February 12, 1919 a friend of the family, Franklin Spencer Edmonds, wired the Woodwards that through the dedicated efforts of Pierre Marinovitch he had located the charred remains of Houston's Spad VII about 50 yards off the road from Montdidier and found the engine plate bearing the serial 10.828. *Caporal* Woodward carried no identification on him during patrols and it was evident that he had been shot down in flames. The Germans, unable to ascertain who he was, had buried him in a nearby shell hole, marking the site with a small cross made from two burnt pieces of the plane.[18] His remains were subsequently reinterred in the American cemetery at Suresnes.

Woodward's loss late in the day narrows down the many German Spad claims on April 1 to that of *Offizierstellvertreter* Johannes Klein of *Jasta* 15. He was credited with a Spad southeast of Montdidier at 1810 hours for his seventh of a wartime total of 11 victories.

Above: *Leutnant* Ulrich Neckel, a 30-victory German ace who, while with *Jasta* 12, claimed a Spad XIII over Villers-Bretonneux on April 12, 1918, probably corresponding to the loss of *MdL* Roger Busch of Spa.94, who was taken prisoner. (Jon Guttman)

On April 7, GC.18 and GC.19 moved to a new aerodrome at Pierrefonds, near Compiègne. Two days later, *Lieutenant* Parizet was transferred to Spa.31. He would only serve with that unit for a month before being invalided to hospital on May 10.

On April 10 Pierre Chan's older brother joined him at Spa.94. Born in Paris on September 23, 1891, Henri Ernest Marcel Chan had risen to the rank of *adjutant* in the infantry by the time he transferred to aviation and earned his military flier's brevet.

On April 11 Bonnet and Marinovitch were on patrol when the former had to force land at the old airfield at Tartiers, northwest of Soissons—only to run into a shell hole. "Marino" returned to Pierrefonds, only to suffer another of his own notorious landing accidents, but then again, so did Martenot

Above: *Lieutenant* René Béchon in his Spad XIII No.IX of Spa.155 at Villeneuve-les-Vertus in March 1918. After sharing the credit in Martenot's victory on April 1, he was shot down later that same day and spent the rest of the war at Neuberg prisoner of war camp. (Christophe Cony Collection)

Above: Airmen line up at Villeneuve-les-Vertus to be made *Chevaliers de la Légion d'Honneur* by *Général de Brigade* Maurice Duval on September 1, 1918. From left, *Capitaine* Hubert de Geffrier, CO, Br.107 (116 bombing missions, one victory), *Sous-Lieutenant* Louis Risacher, Spa.159, *Sous-Lieutenant* René Doucinelle, Spa.48, *Sous-Lieutenant* André Martenot de Cordoux, Spa.94, and *Lieutenant* André Chalus (observer, one victory). (Christophe Cony Collection)

Right & Facing Page, Bottom: French personnel examine the remains of the Albatros D.V shot down by Martenot near Montdidier on April 1, 1918. The tailskid fairing seems to be white, suggesting that the plane is from *Jasta* 17, which lost *Leutnant* Richard Grüter that day. (Christophe Cony Collection)

Above: Germans examine the dismantled Blériot-built Spad XIII of *Adjutant* Henri Chan, brought down in their lines, probably by *Leutnant* Josef Veltjens of *Jasta* 15, and taken prisoner on May 10, 1918. (Greg VanWyngarden)

that day. Neither pilot was hurt, and at 0500 the next morning Bonnet and three mechanics went out to remove the machine gun and instruments from his damaged Spad VII and dismantle it for transport to the *escadrille* for repairs. That would be at a new base, because orders came in for *Escadre de Combat* 1 to move again, this time at Montagne-Fayel, west of Amiens.

Spa.94 suffered another loss during a fierce dogfight over Villers-Bretonneux on April 12, during which the patrol probably downed an enemy scout, but *MdL* Busch, in Spad VII S3191, failed to return. He was brought down and taken prisoner, probably credited as the ninth victory for *Leutnant* Ulrich Neckel of *Jasta* 12, who claimed a Spad XIII over Villers-Bretonneux that day.

In spite of both Houston Woodward's and André Martenot de Cordoux's claims that he was Algerian, more recent information indicates that Roger Auguste Achille Busch had in fact been born in Reims on April 8, 1893. His family later moved to Paris, where he was a lithography designer before the war, also joining the *25e Régiment de Dragons* on October 2, 1913. He transferred to aviation on March 1, 1917, qualified for brevet No. 7,211 at Avord on June 26, and after advanced training joined N.94 on February 11, 1918. After his capture Busch was sent to Karlsrühe prison camp and subsequently transferred to Puchheim before his liberation at the end of the war.

By a strange twist of fate, two new pilots arrived on the same day Busch was lost. Born in Beauvais in the Oise region on July 23, 1893, *Sergent* Maurice Caulier was the son of Léon and Jeanne Vesselin Caulier, and spent much of his childhood in the Château de Montvillargenne at Gouvieux. He was a student when he interrupted his studies to enlist in the 156ème *Régiment d'Infanterie* on November 9, 1913, transferring to the 62ème on August 2, and subsequently serving in the 51ème *Régiment d'Artillerie*. He earned three citations for outstanding service between 1915 and October 2, 1917, when he began pilot training, obtaining brevet No. 10,022 at Istres on October 24. After aerobatic training at Pau and gunnery at Cazaux, he went on to GDE at Bourget-Dugny on January 15, 1918, and was assigned to Spa.94 on April 12. Also arriving that day was *Caporal* Henri Grimouille.

On April 13 Spa.94's personnel completed their move to Montagne-Fayol and by the 19th its aircraft were installed in their assigned Bessoneaux hangars at the edge of the airfield. This had not been achieved without an extraordinary amount of mishaps. On April 17 Bonnet's Spad suffered engine trouble and force landed at Sommelleux, while Pierre Chan, after leaving Pierrefonds, came down at GC.17's aerodrome at Fouquerolles with a breakdown. On the 19th mechanic *Caporal* Bourry left for Roupied where *Caporal* Martin had come down; once his engine was repaired, he returned to Montagne-Fayol the next day. Chan was still stuck at Fouquerolles on the 20th, but finally flew back to rejoin Spa.94 at 1000 hours the next morning. Henri Chan also had his share of engine troubles before bringing in his Spad XIII on the 22nd.

No sooner had it recovered from its epidemic of mechanical failures than Spa.94 was routed by the weather. Four Spads

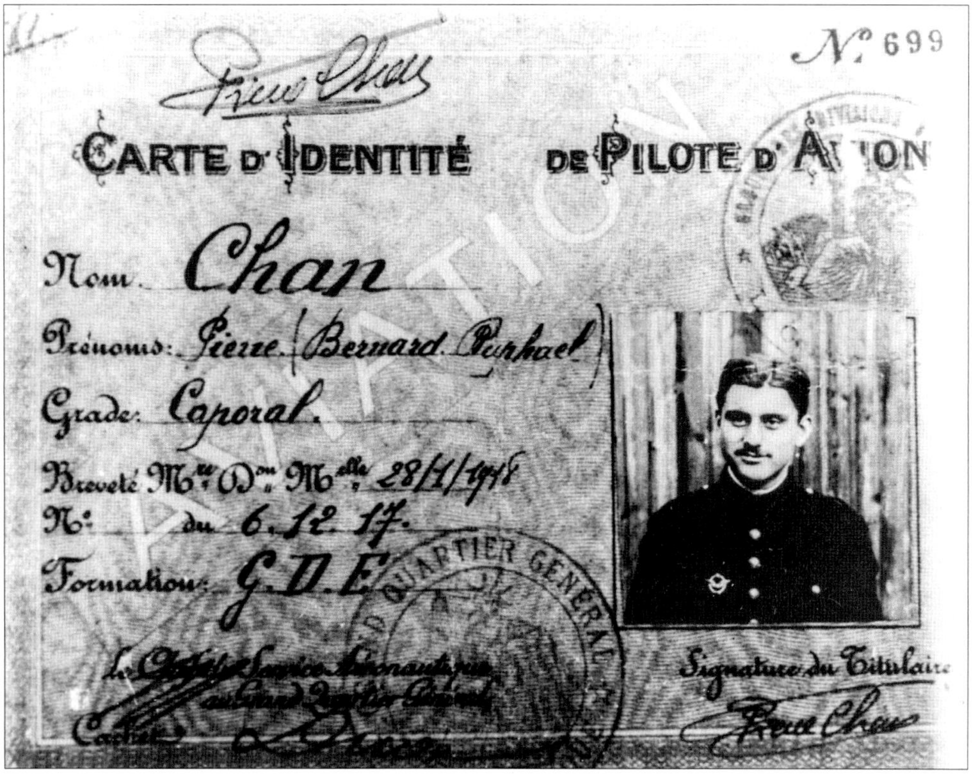

Right: The identification card of *Caporal* Pierre Chan, the younger of two brothers in Spa.94 who were both shot down and taken prisoner in 1918. (Pierre Chan via Jon Guttman)

took off on patrol on the morning of April 24, but soon afterward a pea soup fog descended on the front and it became all the pilots could do just to find their way home. *Adjutant* Edmond Bessières landed on the road near Joue á Crouy, *Brigadier* André Martin landed at Fluy, *Brigadier* Miguel Sainz put down at Boulainvilliers and *Caporal* Eloi Bonnet landed at Estrées-sur-Noye, south of Amiens. Eventually the fog lifted and each of the "downed" pilots found his way back thereafter. To cap off more than a week of frustration, after a general alert flight by GC.18 on April 27, Bonnet telephoned to report that he was down at Frocourt, near Poix, with yet another engine breakdown. A team of mechanics was dispatched forthwith to get his Spad flying again.

On April 29 *Lieutenant* Albert Carbonnel arrived at Spa.94, but a more significant change of personnel occurred on the last day of April when *Lieutenant* Guy de la Rochefordière received orders to transfer from Spa.3 to take over command of Spa.94 from *Capitaine* Pillet, who was reassigned to the GDE on May 2.

Born in Nantes on January 7, 1893, Gonzague Pavret de la Rochefordière was the son of Gabriel Pavret de la Rochefordière and the former Marguerite Marie Hertaux. His career in the *Aéronautique Militaire* had begun in distinguished company indeed, with N.3. Although his spirit was not quite matched by his fighting skills, he managed to score two victories in concert with another N.3 member destined to lead *escadrille* 94 in future. On February 16, 1918, *Sous-Lieutenants* de la Rochefordière and Benjamin Bozon-Verduraz brought down a German two-seater east of Damloup, credited as the first victory for both. Bozon-Verduraz downed a two-seater the next day and one the 20th he and de la Rochefordière shared in another two-seater.

Promising a start as that was, a friend of theirs in N.3, five-victory ace Louis Risacher, was not impressed with either pilot's technique. "Bozon-Verduraz was a very bad pilot," he said in a 1981 interview, "but he had fantastic courage. When he was attacking a German, he would be shouting, '*Pas encore, pas encore*!' He did not maneuver at all! He came right up to the Huns, shot for a few minutes, would not do a half-turn, just went straight on, coming back with his gloves and flying suit shot full of holes. He knew he was a bad shot and a very bad pilot, so he would wait until the last moment, say ten or fifteen meters, before he would shoot.

"*Lieutenant* de la Rochefordière was a very fine fellow, but he disappeared very quickly," Risacher added. "He wasn't a fighter. He was very courageous, but he was not a fighter. When he brought down a German, Bozon-Verduraz kept credit for himself, but he agreed to share credit for two with de la Rochefordière."[19]

On May 3 *Escadre de Chasse* 1 was visited by *Général de Brigade* Maurice Duval, commander of the Bureau of Aeronautics for the *Grand Quartier Général*. From then on the first half of May was marked by further frustration and further losses for Spa.94. After taking off on a patrol at 1430 hours on May 4, Marinovitch attacked a German balloon, but failed to set it on fire, while *Adjutant* Henri Chan's engine suffered a pressure malfunction, though he was able to nurse it back to the aerodrome. A late patrol on May 5 was undone by bad weather, but its members managed to return safely just as night was falling. *Brigadier* Sainz was flying a patrol with aircraft from Spa.48 on June 9 when he ran out of gas and force landed

Above: Henri Chan was probably shot down by *Jasta* 15's *Leutnant* Josef Veltjens. Here Veltjens is shown in his Albatros D.V, not a Pfalz D.IIIa as identified by the French pilots. (via Aaron Weaver)

at Bovelles at 1520 hours. Mechanics brought the necessary fuel and he returned to the *escadrille* without further incident.

At 1430 hours on May 10 *Lieutenant* de la Rochefordière led *Adjutant* Chan and *Caporal* Martin in an attack on two German two-seaters over Avre, but they were in turn bounced by eight Pfalz D.IIIas. By the time the French disengaged, Martin's plane was damaged and Chan went missing again. It was later learned that he had been taken prisoner. The only Spad credited to a German fighter pilot that day went to *Leutnant* Josef Veltjens of *Jasta* 15.

Elsewhere that same day, Marinovitch and Martenot took on a flight of six two-seaters escorted by seven Pfalz D.IIIas over Moreuil. Although the two Frenchmen managed to emerge from the scrap intact, they failed to score any confirmed successes.

"I have heard of the reflections of some German airmen concerning the attitude of French pilots in combat, that we were not as aggressive or 'sporting' as our British allies," Martenot commented. "I don't see how they can generalize like that about anything as individual as air combat of World War I. To me, the ability of the enemy's aircraft was entirely a matter of circumstances. Better than a discussion on the subject, I'll give you the integral text of two of my citations, confirming to you that in 1917 and particularly in 1918, I readily attacked, alone, German patrols whose strength came to about a score of machines, who after some maneuvering abandoned the combat without having shot me down or even hit me, seeing as I am still here:

"'Order of the Army No. 7.851 (13 June 1918).

"'Brilliant fighter pilot, each day provoking the admiration of his comrades by his initiative and his spirit.

"'On several occasions, attacked patrols of ten to fifteen enemy planes and thanks to his sang-froid and his audacity, was able to emerge unhurt from very difficult combats. Just downed his fourth enemy plane.

"'Order of the Army No. 11.513 D (15 November 1918).

"'Officer of the elite, having the highest concept of duty, animated by a courage and initiative above all praise and combining the qualities of fighter pilot of great valor with that of a painstaking observer.

"'On 6 and 12 October, operated a precise survey for our infantry, despite unfavorable atmospheric conditions and gave very important information on the enemy's position.'"

On May 14 *Escadre de Combat* 1 was attached to an even larger combat organization, the 1*ère Division Aérienne*, commanded by *Général de Brigade* Maurice Duval. On that same day *Sergent* Auguste Cornille left Spa.94 due to illness, and Marinovitch attacked an enemy two-seater north of Moreuil. Although he sent it diving steeply toward German lines, there was no reason to believe that it was damaged enough even to rate as a "probable." The next day, however, his luck changed for the better, as recalled by Martenot:

"On May 15 *Adjutant* Marinovitch shot down a German machine, not a two-seater, but a multiplace for long reconnaissance of a type unknown to us, in the area of Esserteaux, south of Amiens, having three officers aboard whom we took prisoner. I don't recall the name of one of them. The two others were Lieutenants Fricke and Prince von Bülow. Before they were taken prisoner, they had had time to render their machine useless."

On May 17 *Adjutant* Ondet attacked a balloon, but again it failed to ignite. The following day saw *Sergent* Martin join two Spads from another unit in a scrap with a *Staffel* of Pfalz D.IIIas, but he returned with nothing to show for his efforts but several hits in his own plane that shattered some fuselage longerons. His may have been the Spad claimed over Montdidier at 1150 hours by *Lt.d.R.* Walter Balzar of Royal Saxon *Jasta* 44 – a victory that was not confirmed. Martenot came under anti-aircraft fire on the 18th and returned with shell holes in his upper wing and radiator.

On May 19 Marinovitch teamed up with *Sous-Lieutenant* Henri Fleureaux of Spa.95 to send a Rumpler two-seater crashing south of Moreuil; its observer, *Leutnant* Kurt Riege of *Fl.Abt. 241(A)*, was killed. The next day Martenot dueled a German fighter and sent it diving out of the fight, leaving behind a trail of black smoke, although it was not confirmed as an official victory.

Jagdgeschwader II: Spa.94's Blue-Tailed Nemesis

Above: Fokker D.VII fighters at Chéry-les-Pouilly in August 1918, with the Prussian blue fuselages and red noses of *Jasta* 15, one of the four *Jagdstaffeln* comprising *Jagdgeschwader* II. Individual markings include, from left, the Uhlan lance of *Vizefeldwebel* Gustav Klaudat (six victories), the three white feathers of *Leutnant* Joachim von Ziegesar (three victories), the "4th Dragoons" monogram of *Leutnant* Oliver von Beaulieu-Marconnay (25 victories), and the 17th Braunschweig "Death's Head" Hussars coat of arms of *Leutnant* Georg von Hantelmann (25 victories).

From its formation in June 1917, *Rittmeister* Manfred von Richthofen's permanent grouping of *Jagdstaffeln* 4, 6, 10 and 11 into *Jagdgeschwader* I, a fighter wing capable of rapid transit to any sector of the Western Front in need of local air superiority, had been stunningly successful against the Royal Flying Corps. Encouraged by that, the commander of the *Luftstreitkräfte*, *Generalleutnant* Ernst von Höppner, authorized the organization of two more such units and on February 2, 1918, *Jagdgeschwader* II was formed, combining *Jastas* 12, 13, 15 and 19 under the overall command of *Hauptmann* Adolf von Tutschek, while *Jagdgeschwader* III under *Oberleutnant* Bruno Loerzer combined *Jastas* Boelcke, 26, 27 and 36. All three *Jagdgeschwader* conducted limited operations while phasing in new stocks of Fokker Dr.I triplanes and preparing for the great spring offensive scheduled for March 21. Before then, however, von Tutschek downed four Allied aircraft, raising his score to 27, but was killed in action on March 15, less than a week before the *Kaiserschlacht* was to commence.

Von Tutschek's replacement, *Hauptmann* Rudolf Berthold, was fanatical, fearless and iron-willed, and made his mark on *JG* II virtually upon arrival. For starters, he insisted on bringing all the aircraft and personnel from his former command, *Jasta* 18, into the new wing, resulting in a unique double change in designation: Berthold's *Jasta* 18 became *Jasta* 15, while *Jasta* 15, under *Leutnant* August Raben, resumed independent operations as *Jasta* 18. With the men and machines, the former *Jasta* 18 arrived with its Berthold-inspired fuselage colors, based on the Prussian blue dress uniform, with red noses based on the cuffs of his old regiment. Berthold went on to make the blue fuselage a common theme for the entire *Geschwader*, with *Jasta* 12's noses being white – the color of the spinners on its black aircraft up until then – *Jasta* 13's green and *Jasta* 19 transferring the yellow from its tails to its cowlings.

Berthold was a demanding leader, though he expected no more from his men than he did from himself. He had been seriously wounded in his right arm on October 10, 1917, leaving him with a withered, nearly useless right hand and pain that still plagued him when he took over *JG* II, although he regarded the prospect of not being able to lead by example just as agonizing. When he got the chance to fly a newly arrived Fokker D.VII in May 1918, however, he appraised it enthusiastically: "Flies very comfortably. Above all, the controls are so light that I can even handle them with my right arm." The Fokker allowed him to open Operation Blücher at the head of his men on May 28, and at 1115 hours that morning he inaugurated his return to combat by shooting down a Breguet 14 over Crouy for his 29th victory. He would score another 15 by mid-August while personally imbuing *JG* II with an aggressiveness that made his "Blue Tails" a force to be respected by the French, and later the Americans, as much the "Richthofen Circus" was by the British.

Escadrille N.94 had already clashed with *Jasta* 19 before its incorporation into *JG* II. It would encounter *JG* II many times throughout 1918, and more than one of its pilots would end his fighting career in a prison camp or a shallow grave as a result. Confusion over confirmations in the war's last weeks results in some discrepancy as to the total victories claimed by *JG* II, varying from 339 to 348. It paid for those successes with 18 pilots killed or mortally wounded in action, another four killed in air accidents, eight taken prisoner and eight wounded in action or injured in crashes.

Above: Pierre Chan at his home in Le Mans in 1979. (Jon Guttman)

On May 28 the German offensive, which had finally stalled along the Somme River, commenced a new phase with Operation Blücher, when the 7., 18. and 1. *Armees* stormed southward along the Chemin des Dames, smashing through General Duchène's *VIème Armée*, as well as the *Xème* and *Vème armées*. West of the Oise, a violent struggle erupted around Matz, as the Germans tried to force a direct route to Paris. On May 29, GC.18 moved to Nantreuil-le-Hardouin, to add its air support to the fight, and commenced operations over the Oise the next day.

"On May 31," Martenot noted, "'Marino' shot down a two-seater over Villers-Cotterets. On June 1, in an individual combat, I made a diving flank attack on a formation of Fokker triplanes and shot one of them down near Parcy. On June 5, with Marinovitch, I was credited with a two-seater over la Ferté-Millon."

Marinovitch's two-seater victim on May 31 was apparently from *Fl.Abt.(A)* 264. Its pilot, *Unteroffizier* Hippolyt Kaminski, was killed, while the observer, *Leutnant* Bake, was taken prisoner. In addition, the "*Benjamin des As*" claimed a Fokker Dr.I over Amblezy that was counted as a "probable." Also initially listed as a "probable," Martenot's June 1 claim was later confirmed and was probably *Lt.d.R.* Rudolf Rienau from Spa.94's old enemy, *Jasta* 19. Brought down near Longpont-Parcy, Rienau was unhurt in the forced landing and rejoined his *Staffel* the next day. Marinovitch had also claimed two Fokker Dr.Is on June 1, but neither was confirmed.

On June 3, Martenot and Marinovitch encountered 12 to 15 "Tangos" and "Checkerboards" over Villers-Cotterets. A desperate 10-minute dogfight ensued. "In this combat," Marinovitch reported, "I felt the wind of a bullet that passed before my eye and brushed my arm before exiting the fuselage." He took three bullet holes in his Spad, but claimed one of his opponents and the French subsequently found a Fokker D.VII down in their lines, which they confirmed as Marinovitch's 10th victory. The nearest recorded German loss that day, however, was *Leutnant* Baum of *Jasta* 79b, whose Pfalz D.IIIa was shot down in flames but who parachuted safely to earth, only to be taken prisoner by the French.

Meanwhile, Martenot had four enemy fighters on his tail in a chase that went from 4,500 to 1,200 meters. Decades later Martenot vividly recalled the incident:

"One of the best examples of why I preferred my handy old, well-balanced Spad VII to the XIII can be found in my toughest fight during this period. On one of my lone forays, I attacked seven Germans and shot down the last one in the lower position. It was not confirmed for lack of witnesses, but the German pilot fell out when his plane flipped upside-down. The leader turned on me and I found myself in the middle of a real free-for-all. I would get under one enemy plane to use him as a 'protector' – the others couldn't shoot at me because that other Boche would be in their way. My Spad was suffering from a faulty motor, but it was outmaneuvering all of them. I couldn't have done that in a XIII. During a turning engagement, one German was so frustrated that he shook his fist at me. About this time, however, three upsetting things happened. Nine more Germans came down on me. While looping, my carburetor closed, choking the engine, forcing me to dive with nine Boches on my tail. Then, when I had enough diving speed I turned at the Germans, fired... and my gun jammed! Nothing left to do then but glide and try to outmaneuver the enemy. I cut through the patrol, breaking it up, then dove again with fifteen converging on my tail, maneuvering desperately as I made for French lines. The pursuit came down from 3,000 to 500 meters altitude, by which time only four were still following me. At that point my motor revived again, and again I turned up and broke them up by flying through them. Then I escaped. It had been a most difficult fight. Yet when my mechanic looked my plane over, he found not one bullet hole in it."

Marinovitch and Martenot were not the only Spa.94 members who were finding action aplenty at that time. On June 4, *Caporal* Henri Grimouille was part of a patrol that caught an LVG east of Villers-Cotterets and chased it back over the lines, during which the German observer put a burst of machine-gun fire through Grimouille's stabilizer and two holes in his upper right wing. The next day Grimouille's patrol was jumped by two enemy fighters, whose fire put holes in his Spad and severed three cables. Grimouille managed to bring his stricken plane down near Vauciennes and, although

Above: Pierre Chan was probably shot down by *Jasta* 27's commander, *Oberleutnant* Hermann Göring, on June 9, 1918. Here Göring, in front of his white-nosed Triplane, briefs his pilots; his Fokker D.VII, marked with white nose and tail, is at left. (via Aaron Weaver)

he turned over on landing, he emerged unhurt. The nearest corresponding German claim was made on a Spad over Chézy-en-Orxois at 2025 hours by *Leutnant* Hans Kirschstein of *Jasta* 6, a component of *Jagdgeschwader* I—the "Flying Circus" of 80-victory ace Manfred *Freiherr* von Richthofen, which since his death on April 21 was commanded by the man the "Red Baron" had named as his successor, *Hauptmann* Wilhelm Reinhard. Another plane in Grimouille's patrol must have at least given the impression that it was in trouble, because a second Spad was credited at the same time over nearby La Ferté-en-Milon by another *Jasta* 6 ace, *Leutnant* Johann Janzen, his 12th and final victory before being himself brought down and taken prisoner three days later.

On June 5 Martenot and Marinovitch were credited with downing a two-seater. This brought Martenot into the ranks of the aces with five official victories to his credit, while raising Marinovitch's score to 11.

It is probably during June 1918 that Spa.94 replaced its composite insignia of the crab and "La République" with Martenot's grim reaper emblem. Martenot claimed it had been chosen by popular vote: "Seeing he could not meet with any enthusiasm for his crab, the captain changed his mind and told all pilots that each should make a proposal and we would draw lots for our choice. Thus '*La Morte Qui Fauche*', begun as a personal emblem, became the *escadrille* insignia, and would survive after the Armistice in active formations, grouped with Spa.62's fighting cock, forming the *Escadron D/I Morvan* for some years, then disbanded and finally re-established at the base at Dijon where it remains to the present day [1982]."

Notwithstanding Martenot's claim of *Capitaine* Pillet relenting, a photograph of Henri Chan's Spad XIII in enemy hands makes it clear that the older insignias were still in use as of May 10, 1918. Pierre Chan recognized the grim reaper as having been on his Spad VII in June, however, and it was certainly in use by July.

June 7 saw *Lieutenant* de la Rochfordière, *S/Lt.* Martenot, *MdL* Maurice Caulier and *Cpl.* Jean-Marie Martre team up to attack a German two-seater, resulting in a "probable" victory. Martre, who had only been in Spa.94 since May 25, was hit and wounded in the fight, and his disabled Spad came down in German lines. Apparently Royal Saxon *Jasta* 24 joined the combat, because its commander, *Leutnant in der Reserve* Heinrich Kroll, claimed to have shot down a Spad VII over the Bois de Laigue for his 25th victory and that subsequently a kite balloon crew of *Ballonzug* 83 confirmed that its pilot had been captured west of the forest at Le Plessis-Brion. The spirited newcomer did not remain a prisoner for long, however – when an opportunity presented itself, he escaped, made his way across the lines and rejoined the *escadrille.* The next day, *Escadre de Combat* 1's area of operations changed to the III*ème Armée* sector.[20]

"On June 9," Martenot said, "We flew two separate patrols. In the Maincamp-St Paul au Bois area, *Adjutant* Marinovitch forced down an enemy two-seater over its own field (confirmed as his 12th victory) and returned with three bullets and a shell-burst in his machine.

"I led the other patrol, which included *Caporal* Pierre Cordier and *Brigadier* Pierre Chan… We went over the lines without any precise goal and had arrived two or three kilometers from them, at around 3,000 meters, when I saw a Salmson reconnaissance plane coming down in a dive, already hit, its engine emitting thick black smoke, at the point of being finished off, and an enemy patrol of five machines approaching their victim rapidly. Unfortunately, we were very badly placed to usefully intervene, being some 300 meters lower. I gunned my engine to try to redress that handicap and signaled my comrades of my intentions for combat. Our direction being head-on, our meeting was imminent; we arrived near the Salmson, which sought to draw nearer to us. I was still about 50 meters lower and to the left of the Germans; it was at that moment that Chan slackened and left our patrol, diving off to the right, in the direction of the distant lines. Not being able to do anything about that, I attacked the enemy in a nose-up turn under their second and third machines; surprised by this unexpected intervention that they could not avoid, the patrol broke up. The leader, still diving, could not recover immediately and made a long bank to come at us again and support his isolated comrades, and in the general melée with us, after two or three passes, it was I who had the superiority in altitude. The five Germans abandoned the combat, diving for their lines without apparent result, but no doubt having taken some hits in the struggle. The Salmson, for its part, had profited by disappearing, and regained its own base after a fashion. Making a half-turn in the direction of our field, I noticed that at the right side of my fuselage a cable had been hit by a bullet and that the sectioned wires were flapping about. That was not the time to apply extra stress; putting my motor at lower revs for less vibration, I rejoined the *escadrille* in a shallow dive without trouble. It was the only enemy bullet I received during all of 1918, despite all my fights. The next day we drove a vehicle with a crew of mechanics to try to find Chan in the area of our combats, but in vain – he had landed his machine between the lines and had been taken prisoner by the Germans."

For several decades, Martenot believed that Chan had deserted his patrol. But Pierre Chan had also survived to present his side of the story during a 1982 interview at his home in Le Mans. Still shaken by the recent loss of his brother, he had indeed held back during Martenot's attack on the first German patrol. Just then, he spotted a second enemy formation of 10 Fokker triplanes emerging from a cloud. He then made the mistake of climbing to engage them – a foolish thing for a Spad to do against one triplane, let alone 10 – and amid a flurry of tracer bullets his engine was hit and stopped. Now Chan had the sense to dive away, but with his motor out of commission, his Spad's advantage was lost and the triplanes followed him to the ground near Longpont.

Chan was probably credited to *Oberleutnant* Hermann Göring, who was leading *Jasta* 27 in a new Fokker D.VII at 0845 that morning when he reported attacking a Spad and that "from 400 meters' altitude it dropped straight down like a stone and hit [the ground] at the northwest corner of Horseshoe Wood south of Corey."[21] It was credited as his 20th victory.

Jumping from his still-rolling plane as bullets flew around him, Chan ran into the woods, became disorientated and spent the next three days wandering and hiding through several artillery bombardments. When advancing German soldiers finally found him, he was too weak from hunger to raise his arms in surrender.

During the passage that eventually led to the prison camp at Lechfeld near Munich, Pierre Chan found Roger Busch and, much to his joy and relief, his brother Henri, whom he had presumed to have been killed. The brothers made several separate attempts to escape, but both failed and spent much time in solitary confinement for their pains. They were finally released after the war.

During the 1920s Pierre Chan was a co-designer of the elegant line of Amilcar lightweight sports cars, which after debuting in 1921 achieved considerable popularity before going into decline in the late 1930s and ceasing production in 1939. During World War II, Pierre and his sister escaped to southern France when the Nazis occupied the north, and ran a chicken farm. Henri Chan fought in the Resistance, but was captured and deported to Flossenburg concentration camp, where he died on October 23, 1944. [22]

On June 10, a Spa.94 patrol caught a Fokker Dr.I attacking a French balloon near Villers-Brettoneux, but failed to bring it down. The *escadrille's* opponent on that occasion may well have been *Leutnant* Willy Gabriel of *Jasta* 11 – Manfred von Richthofen's old red-nosed squadron – who was credited with his fourth victory (of an eventual 11) over a balloon that day, even though there subsequently proved to be no corresponding Allied loss (a rare lapse in balloon homologation, which was usually confirmed only by front-line witnesses who saw the gasbag go up in flames).

Lieutenant de la Rochfordière led the day's patrol at 1100 hours on June 11. At about noon, however, he was attacked from above by two "Albatros D.VIIs" as the squadron diary described them (which was not entirely wrong, given the number of Fokker D.VIIs built under license by Albatros or the Ostdeutsche Albatros Werke). Although the *escadrille* commander dived toward Allied lines, the Fokkers doggedly pursued him and sent his Spad crashing to earth between Méry-la-Bataille and Lataule, where it burst into flames. The remains of de la Rochefordière, who was probably credited to *Leutnant* Kurt Hetze of *Jasta* 13 as his first of an eventual five victories, were found on June 18 and taken to Roissy, where the *escadrille* held his funeral at 1400 hours.

On the same day that de la Rochfordière died, the *Aéronautique Militaire* underwent another organizational change as the French created two air brigades, the *1er* and *2ème Brigades d'Aviation*. *Commandant* Victor Ménard's *Escadre de Combat* 1 and *Commandant* Joseph de Vuillemin's *Escadre de Bombardement* 12 were assigned to the *1er Brigade* under the overall command of *Chef de Bataillon* Louis de Goÿes de Mézeyrac.

On June 12 Caulier was promoted to *maréchal-des-logis*. The next day Spa.94 moved to a new field at Louvre-Epais. At

Above: *Leutnant* Kurt Hetze of *Jasta* 13 and his Fokker D.VII. On June 11, 1918, he shot down and killed *Lieutenant* Gonzague Pavret de la Rochfordière, commanding officer of Spa.94. (Aaron Weaver)

1730 hours on June 15, a Spa.94 patrol engaged a *Staffel* of Fokker D.VIIs, during which *Caporal* Martre was seen to send a Fokker D.VII down to crash near Moulin-sous-Touvent. Soon afterward, however, his flight mates lost sight of him and he did not return to the aerodrome. Martre was later reported killed in action, probably by *Leutnant* Karl Menckhoff of Royal Saxon *Jasta* 72.

Not for the first time, a replacement for a lost Spa.94 pilot would not be long in coming. On June 24 *Lieutenant* Emile Deviterne and *Brigadier* Emile Froussard arrived.

Emile Marie Gustave Deviterne had already made a name for himself, including experience in the illustrious N.38. Born in Cherbourg on March 31, 1889, the son of career officer Louis Victor Octave Deviterne and Marie Clémentine Salzmann, Emile graduated from college in 1909 and promptly followed in his father's footsteps, enlisting in Versailles for a four-year stint in the *7e Régiment de Hussards*. On November 20, 1914 he got a temporary commission as *sous-lieutenant*, which became definitive on December 26, 1915. On May 30, 1916 he transferred into aviation, training as an observer and being assigned that summer to N.38.

Deviterne was wounded in action on July 8, but recovered within the month. On August 8 he was on a reconnaissance with *Lieutenant* Hubert de Fels as his pilot and *S/Lt.* Marcel Henri Burgun providing single-seat escort when they came under attack by Fokker monoplanes of *Armeeoberkommando Fokkerstaffel* 3. Burgin shot down one of the *Eindeckers* in flames, its pilot, *Leutnant* Benno Berneis, dying in German lines near St. Souplice. Beset by two other Fokkers, Deviterne sent one diving down, evidently damaged, into German territory, but the other fighter bore in to hit him twice in the left leg and once in the face near the right eye. De Fels force landed their shot-up Nieuport two-seater in French lines, to be counted as the third victory for AOK 3's commander, *Oberleutnant* Kurt Student.

On August 11 Deviterne was made a *Chevalier de la Légion d'Honneur* and on October 24 he was permanently promoted to *lieutenant*, but a total of four wounds left their scars on him, especially on his left calf, where the muscle atrophied. Nevertheless, on November 29, 1917 he took the remarkable step of applying for pilot training at Chartres and leaving N.38 on December 18. On February 18, 1918 he qualified on a

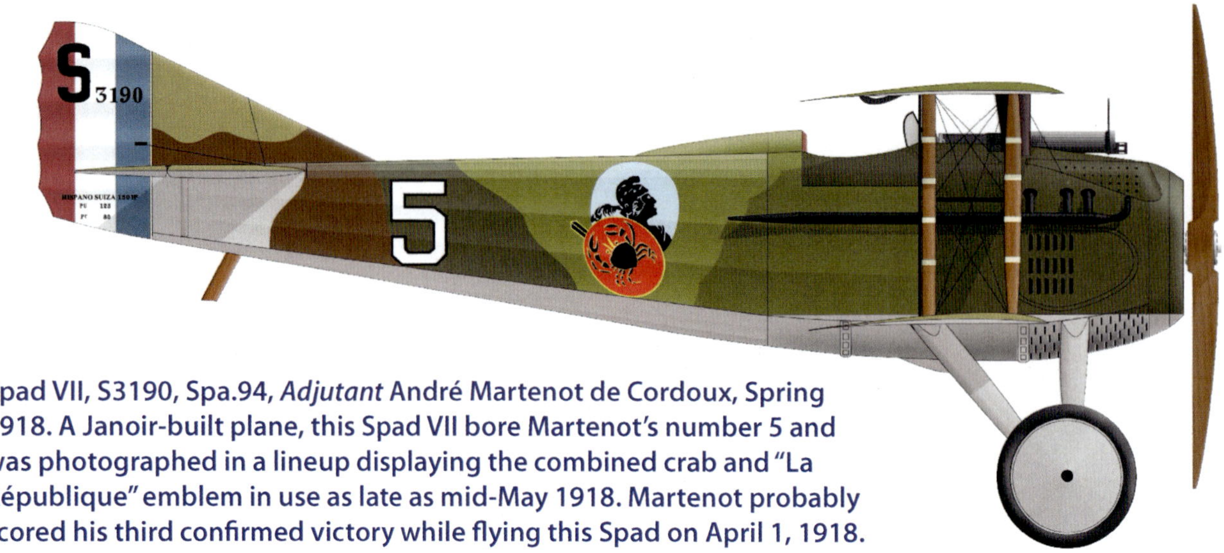

Spad VII, S3190, Spa.94, *Adjutant* André Martenot de Cordoux, Spring 1918. A Janoir-built plane, this Spad VII bore Martenot's number 5 and was photographed in a lineup displaying the combined crab and "La République" emblem in use as late as mid-May 1918. Martenot probably scored his third confirmed victory while flying this Spad on April 1, 1918.

Spad XIII, serial unknown, Spa.94, *Sergent* Maurice Caulier, April 1918. On the day *Sgt*. Caulier reported for duty at Spa.94, April 12, 1918, the ***escadrille*** had lost *MdL* Roger Busch, in Spad VII S3191, probably brought down by *Leutnant* Ulrich Neckel of *Jasta* 12, and taken prisoner. At that time, too, Spad XIIIs were arriving at Spa.94, and Caulier received one…possibly bearing his predecessor's number. If so—and under the circumstances, it seems logical—13 would prove just as unlucky for Caulier as it did for Busch.

Spad XIII, serial unknown, Spa.94, *Adjutant* Henri Chan, May 1918. An early Blériot-built plane, this Spad XIII was brought down on May 10, 1918, Chan being taken prisoner. The best corresponding German claim was by *Leutnant* Josef Veljens of *Jasta* 15.

Farman for military pilot's brevet No. 11,641. He then went for advanced training and aerobatics at Pau from March 4 to April 3, followed by gunnery at Cazaux from April 3 to 29 and, ultimately, his classification as a fighter pilot and assignment to Spa.94.

On June 29 Spa.94's ranks were further swelled by a returning convalescent who was eager to make up for lost time. After Austen Crehore's appendicitis operation, the surgeon had recommended his being reassigned. "He raised such a row at this idea that they sent him home for four months," Marinovitch recalled. "When he got back, we were having a lot of trouble with our Hispano motors, and Crehore was especially unlucky. However, he had many hard fights bringing his machine back full of holes."[23]

By the end of June the German offensive had again stalled, but the enemy had not yet shot his bolt. Moreover, the cost of slowing down the German advance had been heavy for the Allied soldiers—and for the airmen. As all of the warring forces paused to replenish their forward units for a renewed effort, Martenot, among others, paused to appraise his tactics, and those of the *Aéronautique Militaire* in general:

"It is true that I had revolutionary methods of dissuasion contrary to those ordinarily used, as much by French pilots as by the enemy's.

"The American fighter pilots had reason to criticize the method we adopted at a certain time for protection of reconnaissance or bombing planes, which were themselves poorly defended against German pursuit and often victims of their attacks, by tying the scouts close alongside the slower aircraft. In my opinion, this form of protection, though motivated by good intentions, was the error of a badly-informed command, and in any case it was rapidly abolished. The fighters hadn't any rear defense, being easy targets for the attacker, of whom they themselves were occasionally victims without benefit to the plane that was supposed to be protected.

"Like others, I carried out numerous patrols of this sort under orders, but I quickly adopted a personal tactic that proved to be better. Generally, our escort patrols were of three planes, of which I was leader; in case of an attack, I had recommended to my two wingmen that they diverge and leave me alone with the plane to be protected, so as not to hamper me. I slowed down slightly, leaving the field apparently open to the Germans, all the while maneuvering to evade their fire. As they arrived at 200–300 meters from us, I would make a half-turn nose-up and open fire in their direction, with an angle of a few degrees relative to their line of dive. Usually surprised by this attack at close range, the German patrol would literally explode; some would dive to earth, others made a half-turn or sought to come at me again – without result, for in the melée I would have taken the advantage of height over all of them. The protected plane benefited by getting away and I had nothing more to do but try to catch up. I can't say how many patrols of this sort I carried out, but this is certain: That all the protected aircraft returned to their bases as well as myself, generally without having been touched (the action of June 9, 1918 being the exception). This type of combat allowed me to perfect myself and better know the reflexes of the Germans in my sector.

"The Germans weren't the only danger in the air," he continued. "Mistaken identity was a frequent cause of near-tragedy. Once, Sopwith Camels dove on one of our patrols and fired at them. Fortunately, they didn't hit anyone before they saw their mistake, went out of their dive and climbed away. On one occasion, a '*Cigogne*' from GC.12 attacked me. After a short fight, he came alongside and gestured for me to land. We landed together on a road and the pilot personally apologized to me.

"Once I saw a German on the tail of another Spad, and got behind him and fired. The German didn't go down, but the Spad did. Did the German get him, or had I struck him in the act of trying to save him? I may never know – I don't want to know – but from that time on, I never again placed myself directly behind an enemy if he was chasing one of our planes."

Below: Blue-fuselaged, white-nosed Fokker D.VIIs of *Jasta* 12 at Chéry-les-Pomes.

Chapter 5: Going Onto the Offensive

On July 1 Marinovitch, Martin, Sainz and *Sergent* Henri Dussourd engaged a Rumpler two-seater, which Marinovitch sent down to crash near Monnes at 1715 hours. Two days later *Sous-Lieutenant* Benjamin Bozon-Verduraz took command of Spa.94.

Born in Saint Étienne de Cuines (Savoie) on May 29, 1889, Jean Seraphin Benjamin Emmanuel Bozon-Verduraz had entered military service on October 1, 1910, with the *4ème Régiment de Dragons*. He was released from active duty with the rank of *maréchal-des-logis* on September 25, 1912, but was recalled to service when war broke out, as a member of the 9ème *Régiment de Hussards*. Like so many other cavalrymen, he asked for a change of mount and on October 8, 1915 he began flight training at Crotoy. After advance training at Bourget he earned brevet No.2437 on January 19, 1916, and was assigned to C.11 in March. He was injured in an accident on May 12, but during convalescence he was promoted to *adjutant* on November 21, and on May 28, 1917 he was sent to fighter school, after which he was assigned to the famous *escadrille* N.3 on June 15.

Commissioned a *sous-lieutenant* on July 11, Bozon-Verduraz was known for little else in 1917 besides being the last man to see *Capitaine* Georges Guynemer, France's second ranking ace with 53 victories, alive on September 11. On February 16, 1918 he shared his first victory with *Lieutenant* de la Rochefordière, and his tally stood at eight as of May 29. If his friend Louis Risacher's appraisal of him being neither a skillful pilot nor a good shot is true, then Bozon-Verduraz owed as much to luck as he did to his innate fearlessness. Thus far, in any case, it had served him well.

Two days after the new CO's arrival, Spa.94 and the rest of *Escadre de Combat* 1 was ordered to move to Brabant-le-Roi, near the Meuse River, in response to reports of an enemy buildup in that sector. Little activity followed over the next two weeks, but that was to change in the middle of the month.

On July 15 the Germans launched their last offensive along the Marne. On that critical day, *Escadre de Combat* 1 was notified that its aircraft were to cover the fronts occupied by both the IV*ème* and V*ème Armées*, and GC.18 was ordered to the aerodrome at Gourgançon. Few could have suspected it at the time, but the war had truly reached its turning point, as the French – bolstered by units from the American Expeditionary Forces, both on the ground and in the air – gave their all to halt the German onslaught.

Amid some very successful activity by all members of Spa.94 on July 15, the *escadrille's* remaining foreign volunteer, Austen Crehore, managed to steal the limelight. Martenot recalled the day's events:

"A patrol comprising *Lieutenant* Carbonnel, *Adjutant* Marinovitch, *Sergents* Martin, Grimouille and Crehore, and *MdL* Caulier went out and at about 1030 hours and downed three enemy planes in the area of Moronvillers – one enemy plane to Marinovitch, Martin and Grimouille, a Rumpler shared between 'Marino,' Carbonnel and *MdL* Pierre Ducornet of Spa.93, and a two-seater to Crehore. *Sergent* Crehore then got lost in some fog, found himself around Vouzières above an enemy airfield, and downed a plane that was taking off to pursue him. During his return flight, he attacked a *Drachen* and obliged its two observers to jump by parachute."

Guy PAVRET DE LA ROCHEFORDIÈRE

Sous-Lieutenant d'Artillerie. Aviateur

Né à Nantes le 7 janvier 1893, fit ses études aux Enfants Nantais ; élève à S. G. (1910-1912), à l'Ecole Polytechnique (1913-1914).

En août 1914, il fut promu Sous-Lieutenant d'artillerie, et obtint deux citations.

Sur sa demande, il passa dans l'aviation en juillet 1917 ; puis fut envoyé, en février 1918, à l'Escadrille des Cigognes, où dès la première semaine il abattit deux avions, et fut cité deux fois à l'Ordre de l'Armée.

Détaché pour commander l'Escadrille Spad-94, le 11 juin 1918, parti en patrouille volontaire par un temps nuageux, il fut surpris à 300 mètres de hauteur par deux avions ennemis cachés dans les nuages, et fut tué d'une balle en plein cœur. Par un mouvement réflexe sans doute, il dirigea son gouvernail vers nos lignes, où son avion tomba et prit feu aussitôt.

Il obtint ainsi une cinquième citation à l'Ordre de l'Armée, qui lui valut la Croix de la Légion d'Honneur.

Citation O. A.

Véritable chef entraineur d'hommes, ayant pris depuis peu le commandement d'une escadrille, a su lui donner une merveilleuse impulsion en l'entraînant quotidiennement aux combats les plus durs contre les patrouilles ennemies. A été tué en accompagnant avec son audace habituelle l'attaque de notre Infanterie.

Marié au début de mai 1916, il a laissé une petite fille, née en février 1917.

Above: The death notice for Spa.94's second commander, *Lieutenant* Guy Gonzague de la Rochefordiére, who was killed in action on June 11, 1918. (Livre d'Or, Lycée Sainte-Geneviéve)

Above: *Sous-Lieutenant* Benjamin Bozon-Verduraz and *Caporal* Louis Risacher beside a Spad VII of N.3 "Les Cigognes" in July 1917. Bozon-Verduraz took command of Spa.94 following the death of his former squadron mate *Lieutenant* Gonzague de la Rochefordiére. (Louis Risacher Album via Jon Guttman)

Again, Marinovitch had more to say about Crehore's adventures that day. "On the 15th of July," he stated, "He saved another pilot attacked by three Boches, and after a big fight got lost in Germany. He found a German aviation field, and a monoplace started to get him; he didn't give him time, and as the Boche was leaving the ground Crehore jumped on his tail and saw him smash up wonderfully. The clouds were very low and he could not come home, as we often do, by the sun, so he looked on the aviation field and saw where the wind came from; he knew it was blowing from the west, so he made for that direction, following the Forêt de l'Argonne. On his way back he attacked two German sausages and made both observers jump out, but couldn't put them on fire. This Boche was never counted, as he fell too far, and the weather was too bad for him to be seen from our lines. Crehore is one of the finest pilots I know and the best pal I ever had."[24]

On July 18 the German attack had been stopped dead all along the Marne and the Allies wasted no time in going over to the offensive. On that same day, GC.18 completed its move to Gourgançon. Also on the 18th, Bozon-Verduraz was promoted to *lieutenant*, while over the front, Marinovitch probably downed a German fighter and Crehore attacked a balloon with inconclusive results.

The next day saw Marinovitch made a *Chévalier de la Légion d'Honneur*, with the following citation: "A pursuit pilot of very great courage. Gained the admiration of his *escadrille* by facing his very difficult daily combats with exemplary audacity and ardor. Downed his seventh enemy aircraft in two months, and was particularly distinguished during the course of actual operations by reporting his thirteenth victory. *Médaille*

Above: Spad XIII No. 13 of Spa.94, flown by *MdL* Maurice Caulier, who joined the unit on April 12, 1918 – the same day *MdL* Roger Busch was shot down and taken prisoner by *Ltn.* Ulrich Neckel of *Jasta* 12. On August 19 Caulier suffered an identical fate, probably at the hands of *Ltn.* Alfred Fleischer of *Jasta* 17. (SHAA B77.188)

Militaire for feats of war. Six citations." [25]

On July 22, *Capitaine* André Laurent took over command of GC.18 from Sabattier de Vignolle. On the same day, Marinovitch and *MdL* Bessières attacked a Rumpler over Epernay and drove it down in a dive spewing black smoke, while at 1025 hours Martenot made an unsuccessful attack on a two-seater over Champillon.

On July 24 GC.18 moved up to a new airfield at Baye, near Champaubert. Bozon-Verduraz scored a probable victory on August 2, and on August 5 *Sous-Lieutenant* Martenot de Cordoux was made a *Chévalier de la Légion d'Honneur.* His citation read: "Incomparable pursuit pilot, enlivened by a great offensive spirit, veritable example of bravery and skill for his comrades. He reported recently his fourth and fifth victories. One wound. Three citations. *Médaille Militaire* for feats of war."[26]

At about this time in the summer of 1918 copies of the *Saturday Evening Post* began making their way into the hands of Americans on the Western Front, including the Lafayette Flying Corps. Among the featured stories therein the LFC men were appalled to find a series of accounts written by one of their own, Harold E. Wright, describing heroic feats performed under the most arduous and hazardous circumstances. Such braggadocio by an LFC member violated an unwritten rule among his colleagues, who usually downplayed their exploits when interviewed by the press. Far worse, however, was the fact that Wright had performed none of the deeds he described. Among those who knew better was Crehore, for Wright had served in *escadrille* 155, one of 94's neighbors within GC.18.

Born and educated in Brooklyn, New York, Harold Everett Wright enlisted in the *Aéronautique Militaire* on March 20, 1917. From March 25 through September 8 he attended the flying school at Avord, where he got his brevet on the Caudron on July 18, followed by further training at Pau and the GDE before his assignment to N.155.

N.155 had formed at Montdésir on July 12, 1917 and, under the command of *Lieutenant* Edmond François George, had moved successively to La Noblette on the 14th and to Melette on August 6. Wright, arriving on September 11, was its first American, but not its first foreign volunteer. That distinction was held by Géronimo Wilmart de Glymes, born in Buenos Aires, Argentina, on 8 February 1893. His father, Raymond Wilmart de Glymes, was a Belgian disciple of Karl Marx who had been sent to foment revolution in Argentina, only to find the populace unamenable to the ideology. Nevertheless, he decided to live there, practicing law and then teaching as a professor at the University of Buenos Aires.

Wilmart was an engineer when war broke out in July 1914. Perhaps the German invasion of Belgium may have sparked a desire to get involved in the fighting, but whatever the reason, on April 12, 1917 he enlisted in the *1e Régiment Étrangère* and then transferred into aviation. Training at Chartres, Avord and Pau, he obtained military pilot's brevet No. 6741 at Pau on May 31, promotion to *caporal* on June 15, and received French Aero Club brevet No.6715 on August 2. On August 11, he was assigned to N.155, and after a short leave he reported for duty on the 21st.

On September 9, Nieuport 24bis fighters arrived and were assigned to the pilots, Wilmart getting N4467. At that time N.155 had adopted a raven in a brown cape as its unit insignia and Roman numerals for personal identification. Wilmart's plane apparently bore the number X. Wright was assigned Nieuport 24bis N4475 which may, based on his order on the squadron roster, have borne the numeral XII.

Wright's arrival coincided with the disappearance of Georges Guynemer, France's most revered ace. It was an encouraging day for N.155, however, as *Caporaux* Charles Bettend, André Chancerel and Georges Paumier shot an enemy two-seater down in flames near St. Hilaire-le-Petit for the *escadrille*'s first victory. This may have been from Royal Bavarian *Fl. Abt. (A)* 290, which lost *Leutnante der Reserve* Otto Ölsner and Max Nathan Holzinger, killed in action.

There was little further significant activity until 1225 hours on November 1, when two French balloons, stationed near Tahure and Massiges, were attacked and burned by *Leutnant* Pütter of *Jasta* 9, his fourth and fifth victories to date. Later that afternoon *Adjutant* Georges Guiraut, *Caporal* Wright and *Caporal* Wilmart were escorting an Armée Renault reconnaissance plane of AR.60 northeast of Châlons-sur-Marne when they came under attack by six Albatros D.Vs, again from *Jasta* 9, led by *Oberleutnant* Kurt Student. The three Nieuports engaged the Germans, but some got through to the AR which, though hit by them and anti-aircraft fire, managed to force land French lines.

Meanwhile Guiraut and Wilmart were credited with driving an enemy fighter down in German lines and possibly got another, although *Jasta* 9 recorded no casualties. Shortly afterward, though, Wilmart became separated from his comrades and was shot down in flames near Souain, probably by *Leutnant* Otto von Breiten-Landenberg, who claimed a Nieuport south of Navarin Ferme. *Oberleutnant* Student also claimed a Nieuport east of Ripont for his sixth and final victory of the war, but that was probably a misperception of how badly Wright's Nieuport was faring as he dived away; in any case, both he and Guiraut returned to Melette.

Wilmart came down in Allied lines, where his body was recovered and shipped home. Posthumously awarded the *Croix de Guerre* with Palm, he is buried in Recoleta Cemetery in Buenos Aires.

Above: After being ordered home to convalesce after an appendectomy, *Sergent* Austen Crehore rejoined Spa.94 in time for the Third Battle of the Marne and promptly made his resumed presence known over the front in July and August 1918 – as well as excoriating Lafayette Flying Corps colleague Harold Wright in the *Saturday Evening Post.* (History of the Lafayette Flying Corps)

From then on, Harold Wright's career nosedived. He probably had always harbored a timorous streak of which he was unaware until that first fight, climaxing with Wilmart's death, brought it to the fore. In any case, he logged no more than 10 flying hours with no further combats. His squadron mates came to regard him as unreliable and *Lieutenant* George considered his performance unsatisfactory.

On December 3, N.155 got its second American when *Caporal* David W. Guy arrived from GDE. Less than a month later he would transfer to the newly forming N.156. On December 23, Wright departed for the United States, where he applied for a commission in the Army Air Service, only to be rejected because of the French army's negative appraisal. Moreover, he apparently had not given proper notice of his departure, because as of January 19, 1918 on he was listed in French military records as a deserter.

Above: Members of N.155 and visitors from F.60 in September 1917, from left: *Lieutenant* Edmond George, *Caporal* Fernand Barbier, *Caporal* Harold E. Wright, *Adjutant* Jean Desbans (F.60), *Caporal* Géronimo Wilmart de Glymes, *Adjutant* Georges Guiraut, *Caporal* André Chancerel, *Adjutant* Louis Tornare (F.60), *Caporal* Georges Degorce and *Caporal* Paul Montange. (Paul Montage album via Philippe Bentresque)

At that point Wright might well have faded into obscurity, but for his series of *Saturday Evening Post* articles under the collective title of "Aces High." They raised a deluge of protest letters from his LFC comrades, of which one, penned by *Sergent* Crehore on August 4, 1918, made its way into the *Post*'s January 18, 1919 issue:

"Dear Sirs: Due to the fact that I am the only American now left in *Groupe de Combat* 18 of the French Army, I have been requested by *Capitaine* George, as well as by the other French officers and pilots of Harold E. Wright's former *Escadrille*, to write the American press and have the true facts concerning former Pilot Wright's performance brought to light.

"The articles appearing in your widely circulated magazine have caused a good deal of comment among the pilots on the Front here, who knew Harold E. Wright and his history while connected with the French aviation.

"This letter is addressed to you in the hope that you will correct the impressions given by Wright in his articles....you will observe the facts and statements made herein are vouched for by *Capitaine* George, of Wright's own *Escadrille*.

"Besides the many untruths in the articles, the insulting fact is that a pilot, who did not enjoy the confidence and esteem of his comrades, would go to America and, by means of cleverly written articles appearing in a widely circulated magazine, set himself up as the performer and hero of many adventures which were not lived by him at all, but were really achievements of other pilots.

"In the first place 'Sergeant' Wright was a corporal while in the *Escadrille* and left that squadron during December, 1917.

"While with the *Escadrille* he never flew over the enemy lines for more than a total of ten hours and was a participant in but one combat during his ten hours.

"We also declare as false the statement in one morning he had seven fights or combats. No wonder he can't remember the details. Why a fellow with one combat's experience should say he once held the American record for having fought seven in two hours is also far from being clear to us.

"Now he is and apparently had been in America for about three or four months, writing, we admit, interesting articles but ones that contain many large and small absolute falsehoods."[27]

In spite of Crehore's scathing indictment, Wright went on to write fiction after the war. He is believed to have died in Brooklyn in August 1924, although the particulars remain unknown. Since he did – or at least briefly tried to do – his small bit at the front, LFC chroniclers Charles Nordhoff and James Norman Hall thought it only fair to include him in their *History of the Lafayette Flying Corps*, but his entry, aside from the dry particulars of his service, consists solely of their wry summation of his place in LFC annals:

"Harold E. Wright's chief claim to distinction as an airman is due to the series of remarkable flights he made during the summer of 1918, in the Saturday Evening Post Sector. Flying the avion "Remington Typewriter" he had a long and bitter combat with Baron Richthofen, the greatest of German "aces." Richthofen escaped, but Baron Munchausen, the legendary king of ground-flyers, who was hovering at an immense height above the scene of the battle, received a mortal coup from the Wilson machine gun, and fell upward into the blue serene, hoist by his own petard, of which ammunition *Sergent Pilote* Wright had a plentiful supply. So far as is known this was Wright's only official victory."[28]

August 7 saw GC.18 move to Beauvais-Tille, to support the I*ème* and III*ème Armées* as the French counteroffensive continued. At 1715 hours on August 9 *Lieutenant* Bozon-Verduraz and *Sergent* Martin joined a patrol of Spa.48 Spads, which attacked a German balloon near Ligniéres. The French

Right: A photograph of Crehore and Marniovitch taken on July 19, 1918, on which Crehore wrote on the back to his family: "Marry & I. We are of the mind that this is quite good. Conceited boys we are but never the less we've ordered more and I'll send you a clearer one. It was the day Marry was decorated with the Legion of Honor. It's the medal nearest his coat buttons. Notice his *Croix de Guerre avec dix*, rather, 10 palms. Wait for a better picture. I certainly am ashamed of myself up against him – won't wear the lone *Croix*. Austen." Within the month, however, Crehore would make a name for himself and lay the groundwork for his own eventual membership in the Legion of Honor. (Courtesy of Jack Gulick)

were in turn attacked from above by eight Fokkers and in the ensuing dogfight Bozon-Verduraz lost track of Martin. Spa.48, however, reported last seeing him with two Fokkers on his tail before falling in flames over Lignières at 1725. Both Spa.94 members may have been the Spads claimed over Tricot at 1800 hours German time by *Oberleutnant* Rudolf Berthold, commander of JG.II, and *Lieutenant* Oliver von Beaulieu-Marconnay of *Jasta* 15. Bozon-Verduraz's demise was misperceived, but André Martin was killed.

Sergent Crehore also got into trouble that day. He had just returned to the *escadrille* after another short leave due to illness, but aggressively hurled himself into the day's operations and fought several engagements deep in enemy lines. It was a formula for disaster, but Crehore managed to fight his way back with his plane badly shot up. His Spad may have been claimed south of Mourmelon-le-Grand at 1900 hours German time by *Vizefeldwebel* Werner Kettelhack of *Jasta* 63, but nonetheless Crehore survived again and on September 1 he was cited in an

Above: First Lieutenant David Putnam, just transferred to the U.S. Army air Service but still with Spa.38, poses beside *Lieutenant* Georges Madon's red Spad XIII sometime between June 8 and 14, 1918. Flying with N.94, followed by MS.156 and Spa.38, Putnam was a flight leader in the 139th Aero Squadron, USAS, and credited with 13 victories when he was killed in action on September 12, 1918. (Jack Eder Collection via Jon Guttman)

Ordre de l'Armée:

"Pilot of the first order, full of dash, energy and audacity, since his recovery from a long illness returned to his *escadrille* and fought with beautiful spirit, carrying the fight right to the fields of the enemy air arm.

"On August 9, fought some very hard combats in the course of which his machine was riddled with bullets."[29]

Crehore's credited total of two aerial victories was reflected in the *Croix de Guerre* with two palms awarded him. After the war the near-deaf Jerseyite who fought for France because his own country had no use for him would be made a *Chevalier de la Légion d'Honneur*.

On the morning of August 17 Martenot wrote a letter to his parents, describing the past day's events:

"My very dear parents, yesterday evening at the end of the day after a quite hard combat I had against a two-seater, I had a motor malfunction which obliged me to land in a field where there was a big haystack; the plane was completely smashed.

"For some days I won't be flying because I have to find a new machine, then it will have to be put in full working order, which could take about 10 days. Our *escadrille* leader, who left at 0300 hours has still not returned and it is 0700 hours. We are very worried because we know he had some tough fights. We're always fighting harder and I have redoubled my caution."[30]

As things panned out, Martenot had reason to assure his family that he was mixing pugnacity with prudence. After taking off before dawn on August 17, Bozon-Verduraz had opened his account with Spa.94 by driving a two-seater down to crash northeast of Roye, only to be set upon by four Fokker D.VIIs. In contrast to de la Rochefordière's, Bozon-Verduraz's luck held as he succeeded in force-landing his badly damaged Spad in the front lines, and he made it home after Martenot wrote his less-than-sunny letter. On the same day Marinovitch added two more to his tally – a Rumpler and a Fokker D.VII over Roye – but he too paid for it by returning with his Spad riddled by enemy bullets, possibly the Spad credited as shot down over that location to *Leutnant* Veltjens of *Jasta* 15.

Two days later, *MdL* Caulier failed to return, his Spad 13 having been last seen south of Roye at 1750 hours. It was later learned that he had been brought down by *Leutnant* Alfred Fleischer of *Jasta* 17, and taken prisoner.

On August 20, *Escadre de Combat* 1 was designated to cover the III*ème* and VI*ème Armée* fronts, with GC.18 operating from Nanteuil-le-Haudouin. From then until the 22nd, Spa.94 escorted Breguet 14B.2 bombers as they conducted a series of raids against German strategic targets around Trumilly, near Crépy-en-Valois. On the 21st, Martenot and *Adjutant* Ondet came to the aid of some Breguets that had come under attack by enemy fighters and shot down a Fokker five kilometers west of Soissons. Their victim was probably *Vizefeldwebel* Anton Bernhörster of *Jasta* 61, who was killed that day over Montigny, west of Soissons, in Fokker D.VII 2184/18.

On September 6, the entire 1*ére Division Aérienne* was placed at the disposal of Brigadier General William Mitchell, in preparation for the first major offensive of the war to be conducted under the command of AEF General John J. Pershing: A drive to eliminate the German salient around St. Mihiel. As part of that transfer, GC.18 moved to Autrey, southwest of Nancy. Martenot and Spa.94 were about to encounter the U.S. Army.

"In that region," he explained, "the layout of the lines formed a pocket occupied by the Germans, which bulged deep within our territory and remained the same for several years. On September 12, 1918, the attack was decided on and entrusted to the American ground troops, protected by our aviation, of which a patrol of Spa.94 with me as its leader had

Above: *Maréchal des Logis* Miguel Sainz beside his Spad XIII, which bore the number 16. Coming into his stride on September 15, 1918, Sainz quickly scored four victories, three of which were over balloons. (SHAA B77.189)

some engagements with the Germans without result. On the other hand, the fighting on the ground was ferocious, with machine guns and grenades; at the end of the day the Germans disengaged and abandoned a good part of their lines."

Although the American ground offensive got off to a splendid start, and would continue to enjoy success over the five days to follow, September 12 would see the death of a former N.94 member whose loss would be felt among French and Americans alike. Since his entry in the 139th Aero Squadron, USAS, 1st Lt. David Putnam had been credited with three German aircraft by September 12, when he and 1st Lt. Wendel A. Robertson were bounced by seven Fokker D.VIIs of Spa.94's old enemy, *Jasta* 15. Beset by four of the Fokkers, Putnam was credited with bringing down one of them for his 13th victory, while Robertson managed to fight his way clear of his three assailants. Shortly afterward, however, Putnam's Spad XIII S15137, bearing the personal number "0," came down near Limey. Shot twice through the heart, he was credited to *Leutnant* Georg von Hantelmann as his eighth of an eventual 25 victories. Putnam was 19 when his meteoric career came to its end – and so was the German who killed him.

Although a recommendation for the Medal of Honor was not approved, Putnam was posthumously awarded the Distinguished Service Cross by the Americans and made a *Chevalier de la Légion d'Honneur* by the French. He was initially buried in the Forêt de Sebastopol near Toul, alongside Lafayette Escadrille ace of aces Raoul Lufbery, but ten years later both airmen were reinterred in the Lafayette Memorial at Parc de Garches.

"The next day," Martenot continued, "not having to command any patrols, I decided to render a lone account of myself in my Spad VII, and look over the situation in the sector at a low altitude of 1,300 to 1,500 meters. The weather was good, yet a huge cloud of several hundred meters' width hid a part of the blue sky above me to the northeast; for prudence's sake I scanned the sky in all directions before occupying myself with the ground. No French planes were in the vicinity, but coming from the southeast a formation soon defined itself – a group of three Fokker D.VIIs, followed by a fourth at some hundred meters' interval and about 2,500–3,000 meters' height, making their way directly toward the cloud, over which they were clearly going to fly. Without changing my line of flight, acting as if I had noticed nothing, motor lightly reduced, finger on the trigger of my machine gun, I went to arrive at the proximity of the cloud following a trajectory of about 30 degrees to them. The vanguard of D.VIIs entered above the cloud and I lost sight of them; the fourth followed. Some seconds of vain waiting ensued, then the patrol reappeared from the other side of the cloud at the moment that I was going to attack. As I'd lost sight of the fourth, there was a new wait of short duration, then, in front of me at about 100 meters the German pierced the cloud in a strong 70-degree dive to my left. Without waiting further, giving my plane full gas, veering around to the left and climbing I rushed in front of him, aiming at the radiator grille and letting him have a short burst at less than 20 meters, while trying not to ram him. The D.VII did not react and continued his dive to the ground, where he seemed to land normally in a glade below. During this time, his three comrades turned, doubtless to observe what had happened or to attack me anew. Seeing that they could gain no decision and having well-marked the place, I followed the road back to my *escadrille*. Later, having no afternoon duties, I left again in the squadron car, accompanied by Pierre Cordier, a mechanic and a light vehicle driver, to investigate the enemy machine, to find out what happened and if possible to see the traces of the bullets so as to keep an account for corrections to be made in any similar case in future.

"Arriving in the presumed zone where the machine should be found, unable to go further with the car for want of a road and having a wooded area to cross, we left the driver and car at the outskirts and continued on foot without incident, although it was a combat zone with numerous American and German corpses, whom we identified by a helmet or a rifle planted at their side, scattered weapons, smashed machine guns, etc.

"Arriving at the woodside, we saw that we had not been mistaken – we saw the D.VII in the glade, all by itself, with no combatants in the neighborhood. We waited some instants and still nothing happened. Then, we went to the plane, which had, in fact, landed pretty normally. Cordier took some pictures, I examined it at my leisure and my mechanics removed the machine gun. There was still nothing in sight, although we heard some shots close by (we learned soon afterward that the lines held by the Germans were in the woods, a few dozen meters opposite the one in which we arrived).

"The day abated and we had to leave. We backtracked through the woods to get back to the road and our vehicle. We were too late. Visibility in the woods was already bad. We advanced slowly, stumbling over numerous obstacles. After a moment, I told the mechanic to put the machine gun on my shoulder and to go ahead and check the road. He saw nothing there. We had just resumed our trek when suddenly I felt two expressive pokes in the back. 'Stop.' It was some Americans who had been hidden behind the trees near our path, and who had no trouble in surprising us. Cordier, who spoke English fluently, engaged in discussion with them and told me that they had taken us for spies and did not want to desist in thinking so. Good – we asked them to take us to their officer. Agreed, after a short march we arrived before an underground bunker, guarded by armed sentries. Dug in by the Germans and abandoned in the course of the battle; it was several stories deep and spacious – it certainly had more than 150 Americans inside, under the orders of a commander, to whom we were finally led.

"After some palaver, Cordier explained that not only weren't we spies, but French pilots, that it was I who had downed the German fighter that morning and during their attack had protected their progress while leading my patrol. There followed an instant of delirium among the young men present. After calm had been re-established, the commandant explained to us that the German pilot had been wounded and had been evacuated in an ambulance. Taking a memo paper, he drew up at my wishes a declaration of the facts for confirmation. Then, wishing us good luck, he sent us back to our vehicle escorted by four men. It had been prominent enough of a day to keep a memory of."

Although Martenot's recollections seem vivid enough, there is no recorded evidence of a German fighter going down in Allied lines on September 13. *Leutnant* Eugen Kelber of *Jasta* 12 was killed, but his Fokker was reportedly shot down in flames near Longeville, possibly by *MdL* Paul Montange, *Sgt.* Jean Pustienne and *Cpl.* Anatole Chartier of Spa.155. *Leutnant* Rudolf Rienau of *Jasta* 19 was shot down by Spads of the 13th Aero Squadron, but survived by using his parachute and landed not far from the wreck of his plane. Both Fokkers went down in German lines. On the other hand, it is possible that there was a date discrepancy on the part of either the French or the Germans, in which case Martenot's seventh confirmed victory could well have been over *Lt.d.R.* Paul Wolff of *Jasta* 13, who was reportedly wounded and taken prisoner on September 14, his Fokker D.VII, with a green nose and a white arrow on a blue fuselage, being recovered intact by the Americans. *Gefreiter* Anton Kempa of *Jasta* 3 was also brought down and taken prisoner somewhere in the sector that day.

In either case, at the time Martenot's quarry flashed by his sights, he may have been diving to escape a previous attack by two other Spa.94 members. Spa.94's report for September 13, 1918 tersely declared:

"Patrol. One of them composed of *Lieutenant* Laganne, *Lieutenant* Carbonnel and *S/Lieutenant* Martenot de Cordoux gave combat to two D.7s and shot one of them down southeast of Thiaucourt near Viéville."[31] Other records indicate that the Fokker came down near Viéville-en-Haye at 1720 hours. If, however, there was a date discrepancy and the trio actually scored their success on the 14th, and if Wolff was indeed their victim, then he would have represented the first loss inflicted on *JG* II by Spa.94 members since *Jasta* 19's loss to Austen Crehore in January – small retribution enough for the havoc wrought on 94 by that *Geschwader* in the interim months.

At 0930 on the morning of September 15, *MdL* Sainz scored his first confirmed victory – and Spa.94's first over an observation balloon – when he burned a German gasbag over Goin. Three days later, the St. Mihiel offensive was successfully concluded and Spa.94 returned to its old hunting ground, the Champagne. On September 23, the unit moved to Cernon, south of Châlons, from which base it supported the IV*ème Armée's* drive to the Meuse, in conjunction with the U.S. First Army's' advance in the Argonne Forest.

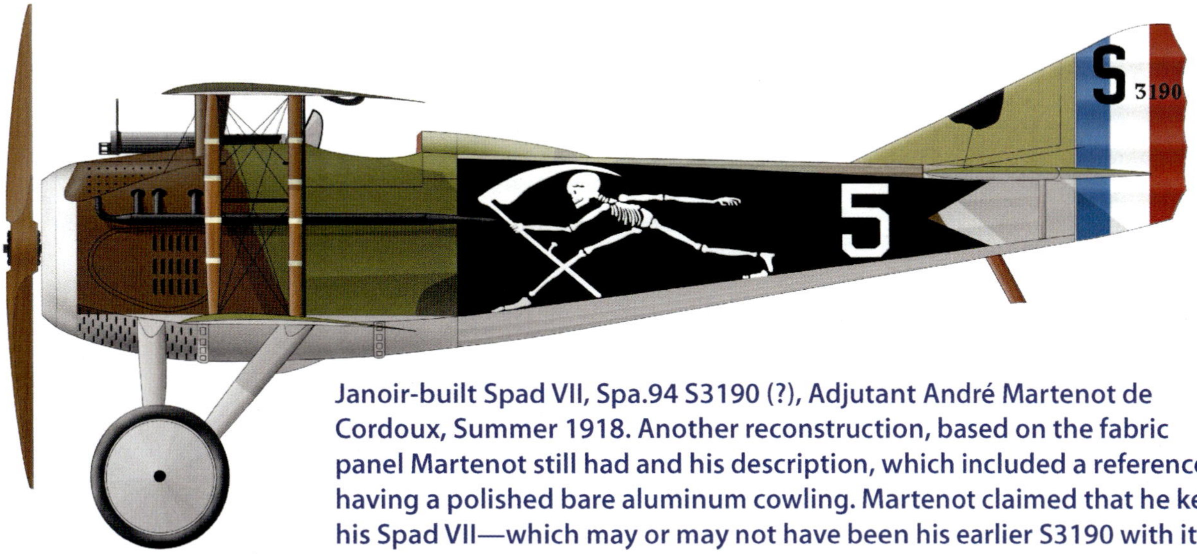

Janoir-built Spad VII, Spa.94 S3190 (?), Adjutant André Martenot de Cordoux, Summer 1918. Another reconstruction, based on the fabric panel Martenot still had and his description, which included a reference to having a polished bare aluminum cowling. Martenot claimed that he kept his Spad VII—which may or may not have been his earlier S3190 with its markings updated—for lone sorties even after he received a Spad XIII for squadron patrols.

Spad VII, serial unknown, Spa.94, *Brigadier* Pierre Chan, Spa.94, June 1918. While interviewed, Chan stated that upon arriving at the *escadrille*, he received *Capitaine* Edouard Pillet's old Spad VII, bearing the number 2 and the legend "Kroknobi" under the cockpit. Somewhat ironically, considering that fact that Chan was Jewish, he was very likely brought down in this plane on June 9, 1918 by *Oberleutnant* Hermann Göring, commander of *Jasta* 27 and future second in the Nazi hierarchy after Adolf Hitler.

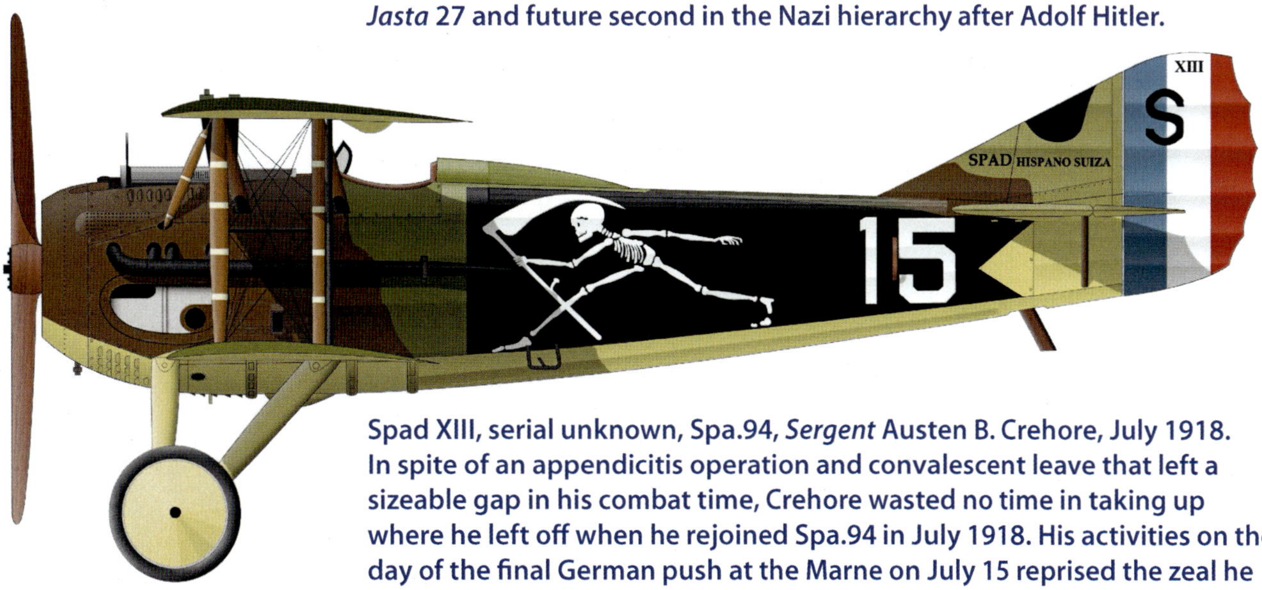

Spad XIII, serial unknown, Spa.94, *Sergent* Austen B. Crehore, July 1918. In spite of an appendicitis operation and convalescent leave that left a sizeable gap in his combat time, Crehore wasted no time in taking up where he left off when he rejoined Spa.94 in July 1918. His activities on the day of the final German push at the Marne on July 15 reprised the zeal he had shown on January 19 and sealed his place in Spa.94 annals.

Chapter 6: Month of Victory

Bozon-Verduraz initiated Spa.94's most successful month by shooting down an enemy plane on October 3. On that same day, Martenot was flying one of his lone expeditions in the area of St. Martin l'Heureux when he jumped two Fokker D.VIIs and fired point-blank into one. According to Martenot, the enemy pilot was hit and the Fokker went down "at least badly damaged." Although Martenot's claim was only counted as a "probable," German records reveal that *Leutnant* Friedrich Höhn, credited with 21 victories and newly-promoted to command of *Jasta* 41, was mortally wounded over St. Martin l'Heureux that day.

On October 4, *Escadre de Combat* 1 was cited in *Ordre de l'Armée* No.10.312 "D" for having destroyed 105 enemy aircraft and 42 balloons since its formation. A Spa.94 patrol tried to add to that by chasing a two-seater from Moronvillers to Pont Faverger, and at 1210 hours Sainz attacked an ascending balloon north of Epaye, but neither attempt achieved success.

At 1130 on the following morning Marinovitch ambushed a Rumpler and two Fokker D.VIIs, and sent one of the latter crashing down east of Challerange. His victim may have been *Leutnant* Arno Benzler, a nine-victory ace and *Staffelführer* of *Jasta* 60, who was wounded in action. Sainz attacked a *Drachen* north of Epoye at 1210 hours that day, but failed to set it on fire.

On October 9, Spa.94 reported: "Patrol over the lines which encountered numerous two-seaters on a large reconnaissance mission. Some combats were fought. *Lieutenant* Bozon-Verduraz and *Lieutenant* Laganne brought down a two-seater in the region of Bathinville. *Brigadier* Froussard attacked two biplaces in 15-minute intervals, one at 1530 hours, the other at 1545. The second fell in a spin and crashed to earth between Machault and Semide. *Sergent* Grimouille attacked a Rumpler two-seater at 1530 without result. *Adjutant* Marinovitch attacked a Rumpler in the region of Semide at 1530. The machine gunner was killed and lay slumped over the side of the fuselage. *Adjutant* Marinovitch suffered a jam but the enemy machine dived steeply into its lines."[32]

Ultimately October 9 proved to be one of Spa.94's most productive days, as victories were credited to Bozon-Verduraz and Laganne, to Marinovitch and to Froussard, whose persistence and skill was further rewarded with a promotion to *maréchal-des-logis* two days later. Grimouille's Rumpler was also credited as his second victory, in concert with two Spa.48 colleagues who finished it off – *Lieutenant* Gilbert Sardier and *Sergent* André Denomaison.

On October 12 *Lieutenant* Martenot led Froussard and *Sergent* Jean Vauzanges on a low-level reconnaissance to observe the advance of French forces around Rethel. In spite of unfavorable weather conditions the trio completed the mission and gathered useful intelligence on the objectives lying ahead of the army. Martenot and Vauzanges than landed at the nearest front-line post to pass on their information to the troops, so they could make timely use of it.

Above: *Sergent* Müller of *Flieger Abteilung* 25 and *Leutnant* Fritz Höhn beside Höhn's Pfalz D.IIIa 8009/17 of *Jasta* 21s. Later appointed *Staffelführer* of *Jasta* 41, Höhn was mortally wounded over St. Martin l'Heureux on October 3, 1918, corresponding to a "probable" claimed over that location by *Sous-Lieutenant* André Martenot de Cordoux. (via Aaron Weaver)

Martenot's activities for the day were not quite over. "On October 12," he claimed, "I forced down a two-seater that landed near the lines east of Tagnon after a very close combat. It was not credited to me; perhaps the Germans recovered the plane. But the pilot and his passenger were killed in the machine by infantry." Martenot's unofficial victims on that occasion may have been *Leutnants* Fritz Guillaume and Bernard Schönleber of Royal Saxon *Fl. Abt.(A)* 226, who were reported in German records as having been killed at Annelles, near Tagnon.

On October 14 Grimouille, Ondet and Sainz downed a Rumpler south of Ecorday. The Spa.94 patrol also went after a *Drachen* near Larey, but intense flak dissuaded the pilots from

Karl Schlegel, Saxony's Forgotten Balloon Buster

Above: *Vizefeldwebel* Karl Schlegel was killed in action on October 27, 1918, probably by *Sous-Lt.* Pierre Marinovitch.

Ironic coincidences abound in wartime and the history of *Escadrille* 94 had its fair share. Among others, David Putnam, the American volunteer who went on to greater things in other units, was 19 years old when he was killed on September 12, 1918, and so was *Leutnant* Georg von Hantelmann, the *Jasta* 15 *Kanone* credited with shooting him down. In an oddly related numerical note, leading Spa.94 ace Pierre Marinovitch survived the war with 22 victories, and among his last victims was a German airman officially credited with the exact same tally.

Born in Wechselburg, Saxony on May 7, 1893, Karl Paul Schlegel had attended military school from 1907 to 1912, after which he served in Royal Sachsenburg Machine-gun Section Nr.19 and subsequently Section Nr 8. From 1914 he served in the 8th Cavalry Division in both France and Russia, earning the Iron Cross 2nd Class that year, the Silver Military St. Heinrich Medal in April 1915, and the Iron Cross 1st class that summer. In the spring of 1917 he quit the trenches for flight training at the *Flieger Erstaz Abteilung* at Altenburg, followed by service in Royal Saxon *FEA* 6. From there he was assigned to *Fl.Abt.* 39, but he was injured in a crash and returned to *FEA* 6 in January 1918.

In May *Unteroffizier* Schlegel was sent to *Kampfeinsitzerstaffel* 1, but five days later he was reassigned to Royal Prussian *Jasta* 45, where his fighting talents began to bloom, starting with the destruction of a French balloon on June 14, another on July 4 and a third two days later. On July 15 he was promoted to *Vizefeldwebel* and celebrated by shooting down two Spads that afternoon. From then on he added steadily to his score, burning balloons on July 19 and 20, downing a Breguet 14 on the 22nd, bringing down another Breguet and destroying a gasbag of the American 1st Balloon Company on the 25th, and finishing the month with Breguets on the 29th and 30th.

A Spad XI fell to Schlegel's guns on August 4, he burned a balloon on the 6th, and scored two double victories over balloons on the 12th and 21st. A Spad on August 29 was rapidly followed by more French balloons eliminated on September 1, 4 and 5.

Schlegel was awarded the Golden Military Merit Cross on October 15, but otherwise it was a fallow month for him – until the 27th, when he attacked a balloon at La Malmaison near Amifontaine at 1540 hours German time, five minutes after which he allegedly engaged 12 French aircraft, possibly driving down a Spad XI. German witnesses thought he had accounted for both aircraft before they lost visual contact, but those last victories were never confirmed because Schlegel never returned to write up a combat report.

sticking around to observe the results.

GC.18 moved to La Noblette, near Châlons in the *IVème Armée* sector on the 18th, and at 1550 that afternoon Marinovitch and Ondet sent an enemy two-seater crashing in German lines. Marinovitch and Carbonnel made an unsuccessful attack on a balloon near Hauteville on October 26, but on the following day Marinovitch was commissioned a *sous-lieutenant.* "Marino" celebrated that career milestone by creating another – at 1405 that afternoon, he shot a Fokker D.VII down in flames over Le Thour. His 20th official victory may have terminated the career of yet another notable German ace, for it coincided with the death of *Vizefeldwebel* Karl Schlegel of *Jasta* 45 at La Malmaison. With a score of 22 victories, 14 of which were over balloons, Schlegel, who had attacked a balloon and a Spad XI near La Malmaison and Amifontaine when his luck ran out, was the last German ace to be killed in action during World War I.

"On October 28," Martenot said, "I got my last enemy, a two-seat fighter in the area of St. Clément. Confirmation was not received, no doubt because of the Armistice."

The Spa.94 combat report for October 28 described that action, as well as a crack by Miguel Sainz at the balloon that

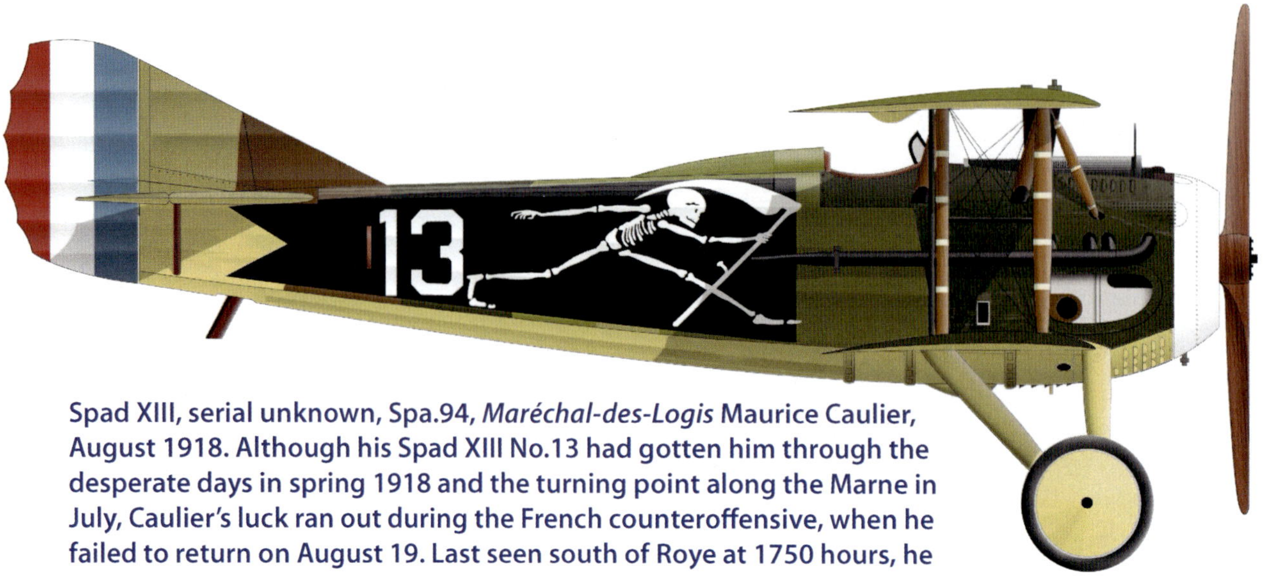

Spad XIII, serial unknown, Spa.94, *Maréchal-des-Logis* Maurice Caulier, August 1918. Although his Spad XIII No.13 had gotten him through the desperate days in spring 1918 and the turning point along the Marne in July, Caulier's luck ran out during the French counteroffensive, when he failed to return on August 19. Last seen south of Roye at 1750 hours, he turned out to have been brought down by *Leutnant* Alfred Fleischer of *Jasta* 17, and taken prisoner.

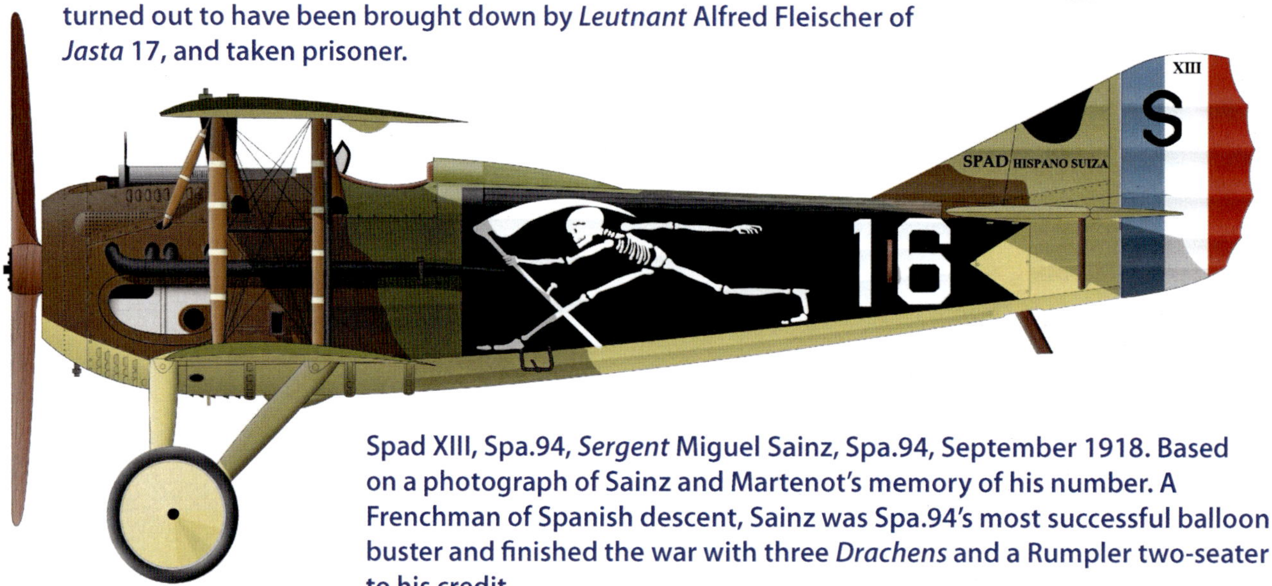

Spad XIII, Spa.94, *Sergent* Miguel Sainz, Spa.94, September 1918. Based on a photograph of Sainz and Martenot's memory of his number. A Frenchman of Spanish descent, Sainz was Spa.94's most successful balloon buster and finished the war with three *Drachens* and a Rumpler two-seater to his credit.

Spad XIII, serial unknown, Spa.94, *Sous-Lieutenant* Pierre Marinovitch, October 1918. While most Spa.94 pilots kept their assigned individual number throughout their time in the *escadrille*, Marinovitch seems to have been the exception to the unofficial rule. Believed to have started out with Nieuport No.14, he was photographed alongside a newly delivered SPAD XIII with the number 7 in April 1918 and was photographed with this Spad XIII late in the war, bearing the number 3. These changes probably reflected Marinovitch's rapid rise up the ranks to a commission and his ascent to Spa.94's leading ace.

Above: *Leutnant* Arno Benzler, *Staffelführer* of *Jasta* 60 and victor over nine Allied aircraft, was wounded on October 5, 1918, possibly by *Adjutant* Pierre Marinovitch. (Greg VanWyngarden)

Marinovitch and Carbonnel had failed to eliminate two days earlier, in greater detail: "Patrol. *Lieutenant* Laganne and *S/ Lieutenant* Martenot de Cordoux attacked a two-seat fighter in the region of St. Clément at 2,500 meters. They forced it to descend into mist that prevented any confirmation, nevertheless it seemed to have been seriously hit. *Lieutenant* Laganne's machine wore numerous consecutive marks of combat. In the report of operations, homologation of the enemy airplane downed on the 27th by *S/Lieutenant* Marinovitch (20th official victory). *Maréchal des Logis* Sainz burned a balloon at Hauteville (3rd machine)."[33]

On the 29th, Carbonnel and Marinovitch tried their luck against the German balloon at Novion-Porcien, but again failed to set it alight. Even worse fortune was suffered that day by *Sergent* Henri Dussourd, whose Spad VII suffered engine trouble, compelling him to force-land near Cernon. Shortly afterward, when he tried to take off and rejoin his *escadrille*, Dussourd's plane crashed and he was killed. His loss was the only dark spot on an otherwise bright month for the Grim Reapers, who had added 10 victories to the squadron tally.

Sainz continued his balloon-busting rampage on the first day of November, swooping down to 500 meters to "roast a sausage" near Macheromesnil, noting that its observer took to his parachute. On the 3rd, Marinovitch and Froussard spotted several German two-seaters strafing French troops and attacked, claiming two probables. Later that day, Marinovitch and Grimouille engaged another two-seater and sent it down to crash near Le Chesne for Spa.94's last confirmed victory. It was also Marinovitch's 21st (not counting the probable on March 26, 1918, which when confirmed would raise his total to 22). In either case, Marinovitch emerged as the *escadrille*'s star performer by a considerable margin.

On November 4, *Adjutant* Emile Picard was wounded in the arm and leg during a fight with a Fokker D.VII before Marinovitch drove it off his tail, only to have to disengage himself when his guns jammed. Spa.94's final combat casualty may have been the Spad claimed east of Reims by *Leutnant* Karl Waldermar Ritscherle of *Jasta* 60, whose victory claim went unconfirmed. November 5 saw another low altitude reconnaissance in the Rethel area, during which the Spa.94 pilots provided French ground forces with useful information on the location of enemy machine gun nests.

Above: French soldiers examine the remains of a Fokker D.VII near Viller-Cotterets, which they identified as Pierre Marinovitch's victim of June 3, 1918. It may have been downed in another fight on another date, however, because the Germans only acknowledged the loss of a Pfalz D.IIIa that day, its pilot, *Leutnant* Baum of *Jasta* 79b, parachuting down to be taken prisoner. (Jean Louis Roba Collection via Christophe Cony)

Left: A somewhat blurry photo of a Spa.94 Spad. (Greg VanWyngarden)

A week later, Germany, facing Allied advances on all fronts and racked by revolution at home, agreed to an armistice. At 1100 hours on the morning of November 11, 1918, the guns on the Western Front ceased firing and aircraft returned to their aerodromes. The war was over.

On November 18 *Lieutenant* Emile Deviterne assumed command from a departing Bozon-Verduraz and the next day Spa.94 moved to Plateau de Malzéville. Citations continued to come in for its pilots, including Bozon-Verduraz, Ondet, Laganne, Grimouille, Sainz and Marinovitch. As a component

Above: *Leutnant* Fritz Pütter (center) when commanding *Jasta* 68 with the other *Jasta* 68 pilots. Variot was probably the ninth victory credited to *Leutnant* Fritz Pütter of *Jasta* 9, who later commanded *Jasta* 68. (Aaron Weaver)

Right: Another view of apparently the same Spa.94 Spad. (Greg VanWyngarden)

of *Escadre de Combat No.1*, Spa.94 was authorized, as of October 15, 1918, to fly the streamer of the *Croix de Guerre avec Palme* from its flagstaff. The squadron was officially credited with 42 enemy aircraft and had produced two aces: Pierre Marinovitch with 22 confirmed victories, and André Martenot de Cordoux, with seven or eight. In addition to those "longtime resident aces," Benjamin Bozon-Verduraz claimed three of his 11 victories while commanding Spa.94 and ace David Putnam began his front-line service in the *escadrille* before scoring all of his 13 victories with other units. Spa.94's wartime losses totaled six pilots killed in action, three taken prisoner and three killed in accidents.

Above: *Sous-Lieutenant* Pierre Marinovitch with what was presumably his last Spad XIII with Spa.94. (Musée de l'Air et l'Espace, Le Bourget)

Above: A lineup of *Jasta* 45 Fokker D.VIIs. The unit's second-ranking ace, *Vizefeldwebel* Karl Schlegel, had 22 victories, including 14 balloons, to his name when he was killed in action, probably by the newly commissioned *Sous-Lieutenant* Pierre Marinovitch, on October 27, 1918. (Greg VanWyngarden)

Above & Below: *Lieutenant* Benjamin Bozon-Verduraz at the end of the war. He scored his 11th and final victory on October 9, 1918. (Steve St. Martin Collection)

Above: The medals on Marinovitch's chest date this photograph as having been taken after late October 1918. (Rich Roberts)

Above: *Sergent* Henri Grimouille shared in four victories with Spa.94, including the *escadrille's* last, when he teamed up with *Sous-Lieutenant* Marinovitch to bring down a German two-seater on November 3, 1918. (SHAA 00.528)

Above: A Spad 13 of Spa.94 in the last months of the war. (SHAA B76.1858)

Above: *Sous-Lieutenant* Pierre Marinovitch poses alongside his Spad in November 1918. (Rich Roberts)

Above: A Spad XIII of Spa.94, resplendent in Martenot's "*La Morte Qui Fauche*" emblem warms up for takeoff in the summer of 1918. (Fernand Chavannes via Steve St. Martin Collection)

Above: From left, *Sergent* Austen Crehore, *Adjutant* Joseph Ondet and *Sous-Lieutenant* Pierre Marinovitch pose together in November 1918. (Rich Roberts)

Above: The remains of a crashed Spad XIII of Spa.94 after the war. This may have been the plane from which Austen Crehore salved the unit insignia as a memento. (Rich Roberts)

Above: The pilot who crashed Spad XIII No.19 evidently emerged from the wreck somewhat the worse for wear but alive. In spite of a slight resemblance to Crehore, Jack Gulick did not recognize this pilot as his grandfather, who would not suffer any airborne mishap of his own until he returned to the States. (Rich Roberts)

Above: Crehore stands beside a Fokker D.VII that has fallen into Allied hands for evaluation after the armistice. (Rich Roberts)

Chapter 7: Après-Guerre

Left: A postwar photograph of a fully bemedaled *Sous-Lieutenant* Pierre Marinovitch, *Escadrille* 94's leading ace and the youngest in the *Aéronautique Militaire*. (Marinovitch was born 1-08-1898. The second youngest was Max Coupillaud (born 8-05-1898), the third youngest was Pierre Ducornet (born 19-04-1898). (SHAA B88.3554)

On November 18, 1918 *Lieutenant* Bozon-Verduraz was invalided to hospital. Command was passed to Lieutenant Emile Deviterne, who led Spa.94 until May 2, 1919. True to his father's occupational choice, Deviterne too would have a long military career, attaining the rank of *capitaine* on June 25, 1923, and commandant on December 15, 1935. On July 2, 1932 he was made an *Officier de la Légion d'Honneur*.

The end of the war was not as happy as it might have been for some of Spa.94's most distinguished members. Austen Crehore, the hard-of-hearing sick case who came back to realize his full potential in the war's last half year, ferried captured German aircraft into France until he was again hospitalized at Sarreguemines on February 6, 1919, and subsequently discharged from French service.

19.
PRICE T

AIR DERBY

AIR RACER CRUSHED
38 MORE FLIERS SPEE
ALBANY ON LAST LE

New York Aviator Suffers Broken Hi
Arm When Plane Crashes Into
Yankees Are Looked Up
Winners of Contest

30 AIRMEN DUE AT ROOSEVEL

Albany watched the sky yesterday while thir
north and west stopped at Quentin Roosevelt Me
national air derby for the Hotel Commodore air
serious accident marred the second day of the
staged.

Today is likely to be the last and greatest da
thirty American planes will be in from the west al
leg of their round trip.

Speeding full tilt into a big elm at the nort
Roosevelt Memorial field yesterday afternoon, on
air after a false landing, an Italian S. V. A. pla
Crehore of New York city, crashed to earth on a s
age. Crehore suffered serious injuries but officials
pital, where he was taken after the accident, said

Five persons standing under the big elm wat
feet away. George Smith of 425 Third street, A
small children, and Joseph Fitzpatrick, a reporte
tree trunk as the big plane's right wing clipped t
dashed to earth below them. Mr. Smith was the f
pilot and dragged him from under the fuselage
gasoline pouring from the tanks might take fire
for fifteen minutes until they were able to forc
spectators which broke from restraint along th
rushed to the scene of the accident.

Skilled air fighters said that it was only by the narrowest margin that Crehore was not killed. He made what observers thought a perfect landing, skipping down over the southeastern edge of the field. His fast plane failed, however, to lose speed as it crossed the field, at the northern extremity was still under good headway. Rather than risk a somersault in the ditch flanking the lot, Crehore again shot his plane into the air. It rose well and skimmed off again, apparently out of danger. Riding in a plane in which observation straight ahead is difficult, Crehore failed to see the elm tree in his path.

German Fok
perfect landi
to Toronto o
ney.

Some of th
the field for
Schroeder, C
likely winne
ing a fast V
York to To
5:34 p. m.,
hour flight
eighty-five
pected to
the race.

an plane; Baron R. D'Orey, French expert
h Austin B. Crehore was seriously injured.
an altitude record, taking the air. 5—Big

PACT'S FATE HANGS
ON SHANTUNG VOTE

Above Left: *Sergent* Austen B. Crehore proudly sports his ribbons as well as the "*Mort qui Fauche*" pin of Spa.94 (Courtesy of Jack Gulick)

Above: A New York newspaper includes photos of Austen Crehore's near-fatal crash outside of Albany in an Ansaldo A.300 on August 26, 1919. (Courtesy of Jack Gulick)

Left: Crehore obtained and preserved Marinovitch's propeller hub after his fatal crash at Evere, Belgium, on October 2, 1919. (Courtesy of Jack Gulick)

Above: The Lafayette Escadrille Memorial in Marnes-la-Coquette, outside Paris. Austen Crehore had much to do with the monument being dedicated to his fellow Lafayette Flying Corps members as well as those of *escadrille* Spa.124. (Courtesy Jack Gulick)

After regaining his health and returning to the United States, Crehore got the backing of a major New York newspaper to enter the First International Air Race from New York to Toronto. On August 26, 1919 he left Mineola, Long Island in a new Italian-built Ansaldo A.300 biplane. With a maximum speed of 120 mph (198 km per hour), Crehore's plane had well outstripped all competitors when he reached the first fuel stop in upstate New York—to find a crowd of enthusiastic onlookers rushing out to meet him on the airfield, utterly ignorant of the danger in which they were putting themselves. Compelled to alter his landing approach to avoid running down his audience, Crehore crashed into a copse of trees and spent more than three months in an Albany hospital. That effectively marked the end of his aviation career. "I believe my mother pressured him to give it up," recalled his daughter, June Gulick.

He soon had another good reason. On October 2, Pierre Marinovitch was giving a flying exhibition to the king and queen of Belgium at the military airfield at Evere, near Brussels, when his plane crashed. Having survived numerous crackups and dogfights, only to die in a postwar accident at age 21, Spa.94's youthful paladin was buried at Père-Lachaise cemetery. Crehore recovered the propeller hub from the wreck and kept it as a memorial to his close friend.

June mentioned one other thing he did: "After Marinovitch's tragic death while performing for the king and queen of Belgium, his dog needed a new home. My father provided it. He brought the dog home with him when he returned to the U.S. Named Patty, I believe, the dog only understood French.

"At the end of the war," June Gulick continued, "after my father returned to the U.S. and after the Albany Air Race, he knew due to his hearing problem, that he wanted to start his own business. In 1919 he formed an insurance agency and an insurance brokerage in New York City and in Westfield, New Jersey, namely Crehore and Richardson, Inc., Brokers Exchange, Inc., and Crehore & Richardson Realty Co. As President he successfully managed them for approximately the next 40 years."

Amid all that activity, Crehore married Katherine Gaffers Dennis on January 14, 1924. They had three children – Katherine June, Austen Jr., who died four months after birth, and Thomas Oliver Crehore.

As demonstrated previously by his letter to the *Saturday Evening Post*, Crehore was a dedicated guardian of the Lafayette Flying Corps' honor and memory. When the French dedicated their *Memorial de l'Escadrille Lafayette* about eight miles outside of Paris on July 4, 1928, he and his wife were both in attendance. "The monument was to honor the memory of the American volunteer pilots of the Lafayette Escadrille and the Lafayette Flying Corps," his daughter June recalled his telling her, "but there was a group who did not want the LFC to be included. In any event, at the dinner the night before the dedication the subject came up again in which a group objected to the inclusion of the LFC pilots. It became a heated discussion until my father objected; in his speech he said that the LFC pilots should not be excluded as they had all fought for the same cause. After that a committee was formed, which included both groups, and they worked to resolve the matter. Ultimately the monument was dedicated to all the American volunteers, whether specifically in Escadrille Lafayette or generally in the LFC."

On January 20, 1933, Crehore was made a *Chevalier de la Légion d'Honneur* by *Commandant* Georges Thenault, air attaché at the French Embassy in Washington, DC, and former commander of N.124 "Lafayette."[34] Crehore later served as

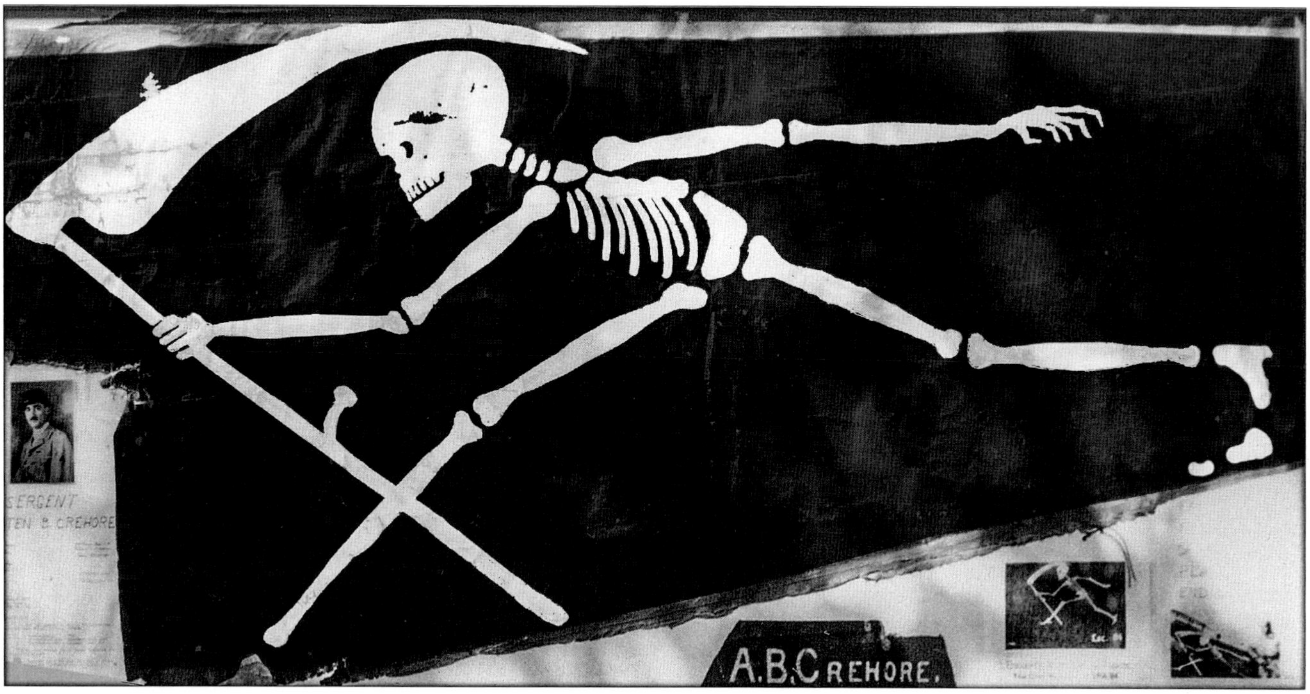

Above: A panel of fabric that Austen Crehore recovered from his last Spad XIII. (Courtesy Katherine B. Hoffman)

president of the Lafayette Flying Corps Association and as director the American Society of the French Legion of Honor Inc. "A last reunion of the Lafayette Flying Corps, which was organized by my father, was held on March 11, 1949 at the old Lafayette Hotel and Restaurant located at University Place and 9th Street in New York City," June Gulick said. "He was a charter member of the Downtown Athletic Club in New York City, and he was chairman of the Finance Committee, Echo Lake Country Club, Westfield, New Jersey.

"Toward the end of the 1950s," June recalled, "with his health declining, he called me up one day to tell me he was about to merge his firm with Eifert, French & Company, Inc., an insurance firm. He would be Vice-President, and his employees' jobs and the services for his customers would continue without interruption. This would mean that no one would lose his or her job, and none of the services he provided to his customers would be cut off, as would have happened had he decided to sell his firm. He was very pleased with the new arrangement. At the time of the merger Crehore and Richardson, Inc., located at 135 William Street, New York City, moved to join Eifert, French & Company, Inc., at their location, 51 East 42nd Street, New York.

"His hobby was wood-working. He spent many hours in his workshop; he remodeled parts of the house, built furniture, and restored antique furniture and antique clocks."

Austen B. Crehore died in a nursing home in Scotch Plains, New Jersey, on August 20, 1962, and is buried at the Fairview Cemetery in Westfield, N.J. His uniform is preserved at the Westfield Historical Society, but his medals, awards and other items are kept at the International Aviation Museum in Colorado Springs, Colorado.

Benjamin Bozon-Verduraz, the finesseless pilot and poor shot who got by on sheer fearlessness and extraordinary luck, accumulated 683 flying hours when he was demobilized from the the *Aéronautique Militaire* in September 1919. Credited with 11 victories, he was also a *Chevalier de la Légion d'Honneur* – elevated to *Officier de la Légion d'Honneur* in July 1920 – with the *Médaille Militaire*, *Croix de Guerre* with eight palms and three stars, and the Russian Cross of St. George. Active in civil aviation in the postwar years, in 1930 he became a member of the *Conseil d'Administration Nationale Aéronautique* and served as part of a commission to promote aviation for tourism. In 1939 he tried to re-enter service in the *Armée de l'Air*, but was turned down as too old. After the fall of France to a new generation of German invaders his health deteriorated rapidly and he died in Chambére on June 21, 1942, aged 53, and was buried at St' Étienne de Cuines.

For his many and varied contributions to his country's victory in World War I, André Martenot de Cordoux was made a *Chevalier de la Légion d'Honneur*, and was also awarded the *Croix de Guerre* with seven palms and star, the *Médaille Militaire* and the Belgian *Croix de Guerre*. After the armistice, he was sent with a team of mechanics to recover abandoned enemy planes. Commandeering ferry pilots as necessity demanded at Metz, Strasbourg, Sarrebourg, Bitche and Sarreguemine airfields, he brought about 350 German aircraft of all types to Le Bourget. On May 5, 1919, he was promoted to reserve infantry lieutenant, effective March 26. On October 25 he retired from active duty as part of the great postwar demobilization process.

Settling down in the small town of Billancourt, Martenot kept his hand in aviation by taking refresher courses at the

Above: Crehore's tunic at the Reeve House, Westfield, N.J., preserved by the Westfield Historical Society. (Courtesy of Jack Gulick)

3ème *Régiment d'Aviation* from August 16 to 30, 1920, at the 2ème *Régiment* from June 15 to 30, 1922, and a third flight training course from June 11 to 25, 1923. Meanwhile, in July 1921 he reorganized the Beauval aerodrome, acquiring five war surplus Nieuports and Caudrons with which he proposed to conduct primary training for a new generation of fliers.

Martenot married 27-year-old Parisienne Germaine Berthe Louise Henrion at Compiègne in September 1925. He pursued his career as a civilian flying instructor between 1930 and 1939, when he returned to the *Armée de l'Air* as commandant at the *Centre d'Instruction de la Chasse* at Chartres, which gave advanced training in Morane-Saulnier 406, Bloch 152 and Curtiss H-75A (American-built P-36 Hawk) fighters. He asked for assignment to an operational squadron when World War II broke out in September 1939, but his request was denied.

During the German invasion in May–June 1940, Martenot had the unhappy task of directing the group's retreat as an element of the 122*ème Bataillon de l'Air* under *Colonel* Pierre Marie Robert Bladinières. The group's aircrews retreated to the Atlantic coast, then to the Mediterranean coast and were about to fly to Algeria when word arrived that another armistice had been signed on June 18 – this time with the Germans as the victors.

Released from service in the *Armée de l'Air*, Martenot retired to Paris and returned to civilian endeavors during the German occupation – endeavors that included those of spy and informer for the French Resistance. "On September 15, 1943," he recalled, "fifteen Boeing B-17s of the U.S. Eighth Air Force bombed the Hispano-Suiza plant in the Paris suburbs (the Bois-Colombes). One bomb fell on my street, the Rue Nouvelle, shattering windows and damaging my house. My wife and daughter hid in nearby trenches as the bombs fell."

With the Allied victory in 1945, Martenot returned to

Above: André Martenot de Cordoux as *Commandant* of the *Centre d'Instruction de Chasse* at Chartres in 1939. (A. Martenot de Cordoux)

military service doing office work with the 2*ème Bureau de l'Armée*, retiring from military affairs in 1946 to work as an engineer and an electrician. As of 1982, he was living in well-earned retirement in Paris' XI *Arrondissement*. At that time, his only remaining squadron-mate, Pierre Chan, was residing in Le Mans. One other old acquaintance also survived: *La Mort Qui Fauche*. Martenot saved a large panel of fabric from the side of his Spad VII, with his grim creation and his numeral "5" behind it. In 1979, he donated it to the *Musée de l'Air et l'Espace* at Meudon, which took steps to preserve and mount the artifact.

In the years that followed my visits to Paris, I lost contact with both Martenot and his daughter Elizabeth. Subsequent attempts to ascertain whether he had died, and if so, when, drew a blank, even from such an authority as the *Service Historique de l'Armée de l'Air*. Then, on November 22, 2014, information finally turned up. André Martenot de Cordoux had died on December 29, 1991, and was buried in the Saint-Ouen cemetery. He was 98.

Documentation aside, the panel from his Spad VII can still be seen on display at the *Musée de l'Air et l'Espace's* modern facility at le Bourget. Gathered with samples of other unit insignias, that black banner with *La Morte Qui Fauche* stands as a monument not only to Martenot, but to all of the pilots and ground personnel of *Escadrille* Spa.94.

Above: Martenot and his daughter, Elizabeth Martenot de Cordoux, at home in Paris in June 1978. (Jon Guttman)

Above: Jon Guttman and André Martenot de Cordoux outside the *Musée de l'Air* at Chalais Meudon in June 1978, after Martenot donated the fabric side panel from his Spad VII to the museum. (André Martenot de Cordoux via Jon Guttman)

Above: *Leutnant* Fritz Höhn prepares for takeoff in his Fokker D.VII while with Royal Saxon *Jasta* 21. On August 25, 1918 he was put in command of *Jasta* 81 and briefly served as acting commander of *Jasta* 60 before taking command of *Jasta* 41 on September 30. Four days later he was mortally wounded over St. Martin l'Heureux, possibly by *Sous-Lieutenant* André Martenot de Cordoux. (Courtesy Aaron Weaver)

Above: The panel from Martenot's Spad VII, showing Spa.94's insignia and his personal number, which he presented to the *Musée de l'Air* at Meudon in 1978, can be seen at upper right among insignias from other famous *escadrilles* at the *Musée de l'Air et l'Espace* at Le Bourget. (Jon Guttman)

Below: Another view of "Les Tangos," the Albatros D.Vs of *Jasta* 19. D.V 2111/17 in the foreground, bearing the initial "M" on the fuselage side was flown by *Vizefeldwebel* Martin Mallmann, whose death north of Beine on January 19, 1918 was jointly credited to Crehore and Marinovitch. (Courtesy Aaron Weaver)

Above: The funeral for *Sous-Lieutenant* Pierre Marinovitch, who crashed to his death on 2 October 1919. (Christophe Cony)

Above: Unarmed Nieuport 17 or 23 wearing early Spa.94 crab insignia. (Jean Pierbattista via Albin Denis)

Facing Page: *Sous-Lieutenant* Pierre Marinovitch looks very much the *"Benjamin des As"* in this formal portrait. He survived the war, but did not live long past his 21st birthday.

Appendix A: Pilots Known To Have Flown With Escadrille Spa 94

Personnel	From	With *Escadrille* 94	To
Commanding Officers:			
Capt PILLET, Edouard		1 Jun 17 – 30 Apr 18	GDE
Lt GONZAGUE PAVRET de la ROCHEFORDIÈRE, Guy	Spa 3	30 Apr 18 – 11 Jun 18	KIA
Note 1		11 Jun 18 – 3 Jul 18	
Lt BOZON-VERDURAZ, Benjamin	Spa 3	3 Jul 18 – 11 Nov 18	
Pilots:			
Adj BARRIOT, Pierre		1 Jun 17 – 1 Jan 18	N 156
S/Lt BELLOC, Raoul		28 Nov 17 – 1 Jan 18	N 156
Adj BESSIERES, Edouard		27 Sep 17 – 28 Aug 18	CIACB
Cpl BONNET, Eloi		10 Dec 17 – 1 May 18	Hosp
Cpl BOSTIERE, Marcel	N 38	1 Jun 17 – 7 Aug 17	CRP
Sgt BUSCH, Roger		18 Feb 18 – 12 Apr 18	POW
Cpl CARBONE, Etienne	GDE	8 Jun 17 – 14 Jun 17	CRP
Lt CARBONEL, Albert		29 Apr 18 – 11 Nov 18	
Brig CASTANEDA de CAMPOS, René	Spa 48	16 Jun 18 – 23 Jun 18	Evac.
Cpl CATHALA, Fernand		16 Aug 18 – 11 Nov 18	
Sgt CAULIER, Maurice		12 Apr 18 – 19 Aug 18	POW
Cpl CHAMPAGNOLE, Emile	GDE	1 Jun 17 – 2 Jul 17	GDE
Adj CHAN, Henri		10 Apr 18 – 10 May 18	POW
Adj CHAN, Pierre	GDE	30 Mar 18 – 9 Jun 18	POW
Cpl CHARPENTIER, Jean		15 Oct 18 – 11 Nov 18	
MdL CHATILLON, Edouard		17 Dec 17 – 1 Jan 18	N 156
Cpl COLLET, Françisque	CIACB	17 Sep 18 – 11 Nov 18	
Lt CONSTANS de SAINT SAUVEUR, Henri	N 38	18 Jul 17 – 10 Nov 17	CO N 157
Adj CORDIER, Pierre	GDE	27 Sep 17 – 5 Oct 18	
Sgt CORNILLE, Auguste	GDE	8 Jan 18 – 14 May 18	Hosp
Sgt CREHORE, Austen B. (American)	GDE	3 Dec 17 – 7 Feb 18	Hosp
Lt DAVITERNE, Emile		24 Jun 18 – 11 Nov 18	
Lt de LAGERIE, François		11 Dec 17 – 20 Mar 18	Hosp
Lt de LAGERIE, François		6 Jun 18 – 10 Jun 18	Maroc
Cpl DELAUNAY, Jean		17 Dec 17 – 1 Jan 18	N 156
Adj DESMOND, Pierre	F 71	1 Jun 17 – 17 Jul 17	Hosp
Sgt DUSSOURD, Henri		19 May 18 – 29 Oct 18	KIAcc
MdL FROUSSARD, Emile		25 Jun 18 – 5 Jul 18	GDE
MdL FROUSSARD, Emile		22 Jul 18 – 11 Nov 18	
Sgt GARLAND, Louis		13 Sep 18 – 11 Nov 18	
Sgt GARNIER, Jean		1 Jun 17 – 1 Jan 18	N 156
MdL GRANGERON, Camille	GDE	21 Oct 17 – 1 Jan 18	N 156
Adj GRIMOUILLE, Henri		12 Apr 18 – 11 Nov 18	
Sgt GROSS, Henri	GDE	3 May 17 – 25 May 17	MIA
Sgt HERBERT, Marcel		9 Dec 17 – 1 Jan 18	N 156
Lt LAGANNE, André	GDE	15 Dec 17 – 11 Nov 18	
S/Lt MARINOVITCH, Pierre	N 38	23 Jun 17 – 11 Nov 18	
Adj MAROC, Eugène		15 Oct 18 – 11 Nov 18	
S/Lt MARTENOT de CORDOUX, André		1 Jun 17 – 11 Nov 18	

Personnel	From	With *Escadrille* 94	To
Sgt MARTIN, André		10 Dec 17 – 9 Aug 18	KIA
Cpl MARTRE, Jean-Marie		25 May 18 – 15 Jun 18	KIA
Sgt MOREAU, Marcel		1 Jun 17 – 4 Oct 17	
Adj ONDET, Jean Joseph	GDE	31 Oct 17 – 8 Nov 18	Evac.
S/Lt OTTAVY, Alexandré		1 Jun 17 – 1 Jan 18	N 156
Lt PARIZET, Georges		1 Jun 17 – 9 Apr 18	Spa 31
Capt PAUMIER, Emile	N 76	11 Nov 17 – 1 Jan 18	CO N 156
Adj PICARD, Emile		15 Aug 18 – 4 Nov 18	WIA
Cpl PUTNAM, David E. (American)		12 Dec 17 – 1 Jan 18	N 156
Sgt RAYER, Léo		1 Jun 17 – 27 Dec 17	KIAcc
Sgt REGIS, Jean		1 Jun 17 – 20 Oct 17	Inj/DOW
Sgt ROYER, René		11 Dec 17 – 1 Jan 18	N 156
Lt CONSTANS de SAINT SAUVEUR, Henri	N 38	18 Jul 17 – 10 Nov 17	CO N 157
Adj SAINZ, Miguel		20 Jan 18 – 11 Nov 18	
Cpl SCHAFFER, Walter J. (American)		16 Dec 17 – 1 Jan 18	N 156
Sgt VAN DEN BOSCHE, Léon		3 Jun 17 – 1 Jan 18	N 156
Adj VARIOT, Henri		1 Jun 17 – 1 Jan 18	N 156
Adj VAUZANGES, Jean		26 Aug 18 – 11 Nov 18	
Cpl WINTER, Wallace (American)		3 Dec 17 – 1 Jan 18	N 156
Cpl WOODWARD, Huston (American)		17 Dec 17 – 1 Apr 18	KIA

Notes:

1. The author has found no reference for who commanded Spa.94 during this period and can only conclude that the XO performed those duties until a permanent replacement could be found (probably Lt André LAGANNE)
2. Personnel roster checked 13 November 2007.

Below: Two MoS..30.E1 trainers display their Le Rhône engines and additional wing bracing cables. These aircraft are not armed.

Appendix B: Spa 94 Pilots Credited with Victories

	Spa 94 Shared	Spa 94 Alone	Spa 94 Total	Other Units	War Total
S/Lt Marinovitch	15	7	22	–	22
Lt Bozon-Verduraz	2	1	3	8(1)	11
S/Lt Martenot de Cordoux	4	3	7	1(2)	8
Sgt Grimouille	1	3	4	–	4
MdL Sainz	3	1	4	–	4
Lt Laganne	–	3	3	–	3
Adj Ondet	–	3	3	–	3
Lt Carbonel	–	2	2	–	2
Sgt Crehore	1	1	2	–	2
Lt Gonzague de la Rochefordière	–	–	–	2(3)	2
Lt Parizet	–	2	2	–	2
Adj Variot	2	–	2	–	2
Brig Froussard	1	–	1	–	1
Cpl Martre	1	–	1	–	1
MdL Bessières	–	1	–	–	1
Cpl Bonnet	–	1	1	–	1
Cpl Woodward	1	–	1	–	1
Sgt Van Den Bosche	1	–	1	–	1
Sgt Martin	–	1	1	–	1
Adj Desmond	–	–	–	1(4)	1

Notes:
(1) Eight with Spa.3.
(2) One with C.28.
(3) Two with Spa.3
(4) One with N.514/F.71

Right: A Spad XIII pilot of Spa.94 whose identity has not been positively ascertained. (Rich Roberts)

Appendix C: Escadrille 94 Combat Log

In the course of researching the victories of *Escadrille* Spa.94 and its two resident aces, Pierre Marinovitch and André Martenot de Cordoux, a surprising number of discrepancies turn up, caused by confusion between confirmed and probable victories, and even the airmen involved in each. Officially, the unit was credited with a total of 42, although there is reason to believe that at least one or two of the "probables" may in fact have been confirmed but not retroactively credited. As mentioned in the text, for example, there is evidence that the unit's "probable" victory of March 26, 1918 was confirmed after the war, affecting the scores of Pierre Marinovitch and two other pilots, and is therefore included here as a "43rd" victory. Spa 94 was credited with 43 victories, 40 aircraft and three balloons. Apologies are offered in advance for any errors in judgment in the course reconciling the inconsistencies to produce this appended list.

Date	Pilot(s)	Victory No. Unit	Victory No. Pilot	Enemy Aircraft	Location	Remarks
13.07.17	*Sgt* Variot	1	1	Albatros C	Mont Haut	*Vfw* Wilhelm Neuhaus, KIA
25.07.17	*Adj* Martenot	2	2	Rumpler C	Mont Cornillet	*Gfr.* Bruno Jakob, *Ltn.* Heinrich Burkert, *Fl Abt(A)* 206, KIA
08.09.17	*S/Lt* Parizet	3	1	Albatros D.V	St. Hilaire-le-Petit	
	MdL Marinovitch		1			
11.09.17	Adj Martenot	–	Prob.	2-seater	W Beine	
20.10.17	*Sgt* Jean Régis crashed in Nieuport 27, later died of injuries at Auxerre.					
01.11.17	*Adj* Variot	4	2	2-seater	Souain	*Ltn* Friedrich Hahn, *Ltn* Karl Mangold, *FlAbt* 247, KIA
05.12.17	*MdL* Marinovitch	5	2	Rumpler C	Mourmelon-Baconnes	
05.12.17	*Sgt* Van Den Bosche	6	1	Albatros C	Somme-Py-Aure	
22.12.17	*MdL* Marinovitch	7	3	Rumpler C	N Pont Faverger	
27.12.17	*MdL* Léo Rayer killed in crash, Sgt Léon Van Den Boche injured in crash.					
01.01.18	*MdL* Marinovitch	8	4	Rumpler C	Beine-Epoye	
03.01.18	*Cpl* Charles Nicholas WIA, died of wounds next day.					
04.01.18	*S/Lt* Parizet	–	Prob.	2-seater	NW Somme-Py	
06.01.18	*Cpl* Woodward	9	1	Albatros D.V	Manre	
19.01.18	*MdL* Marinovitch	10	5	Albatros D.V	Manre-Beine	*Vfw* Martin Mallmann, *Jasta* 19, KIA
	Cpl Crehore		1			
25.01.18	*Lt* Laganne	11	1	Scout	Fontaine-en-Dormois	
	Lt Parizet		2			
	MdL Bessières		1			
	Brig Bonnet		1			
27.01.18	Patrol	–	Prob.	EA		
26.03.18	*Lt* Laganne		5	2-seater	Caurel	
	Lt Parizet		4			
	Adj Marinovitch		22			
	Cpl Martin		1			
01.04.18	*Adj* Martenot	12	3	Albatros D.V	Chaulnes-Montdidier	*Ltn* Richard Grüter, *Jasta* 17, KIA
	Lt Bechon, Spa.155		1			
01.04.18:	*Cpl* Henry Houston Woodward KIA SE Montdidier by Ltn Johannes Klein, Jasta 15.					
12.04.18:	Patrol	–	Prob.	Scout	Villers-Brettoneux	*Jasta* 12, pilot unhurt
12.04.18:	*MdL* Roger Busch POW, Villers-Bretonneux, by *Ltn* Ulrich Neckel of *Jasta* 12.					
10.05.18:	*Adj* Henri Chan POW, Avre, by *Ltn* Josef Veltjens, *Jasta* 15.					
14.05.18	*Adj* Marinovitch	–	Prob.	2-seater	N Moreuil	
15.05.18	*Adj* Marinovitch	13	6	3-seater	Esserteaux	*Ltn* Frick, *Ltn* Prince von Bülow, POW
19.05.18	*Adj* Marinovitch	14	7	Rumpler C	S Moreuil	
	S/Lt Fleureaux, Spa.95	3				
20.05.18	*Adj* Martenot	–	Prob.	Scout		
	S/Lt Fleureaux, Spa.95		3			
31.05.18	*Adj* Marinovitch	15	8	2-seater	Villers-Cotterêts	*FlAbt(A)* 264, *Uffz* Hippolyt Kaminski, KIA, *Ltn* Bake, POW
31.05.18	*Adj* Marinovitch	–	Prob.	Fokker Dr.I	Amblezy	
01.06.18	*Adj* Martenot	16	4	Fokker Dr.I	Parcy	*LtdR* Rudolf Reinau, *Jasta* 19, unhurt
01.06.18	*Adj* Marinovitch	–	Prob.	2 Fokker Dr.Is	Longpont	
03.06.18	*Adj* Marinovitch	17	9	Scout	la Ferté-Millon	In French lines
	Adj Martenot			Prob.		
05.06.18	*Adj* Marinovitch	18	10	2-seater	E Faverolles	
	Adj Martenot		5			
09.06.18	*Adj* Marinovitch	–	–	2-seater	Maincamp-St.Paul-au-Bois	
09.06.18:	*Brig* Pierre Chan POW, Longpont by *Ltn* Hermann Göring, *Jasta* 27.					
10.06.18	Patrol	–	Prob.	Fokker Dr.I	Villers-Brettoneux	*Ltn* Willy Gabriel, *Jasta* 11, unhurt

		Victory No.				
Date	**Pilot(s)**	**Unit**	**Pilot**	**Enemy Aircraft**	**Location**	**Remarks**
11.06.18:	*Lt* Guy Gonzague de la Rochefordiére KIA Montener-Méry Lataule by *Ltn* Kurt Hetze, *Jasta* 13.					
15.06.18	*Cpl* Martre	19	1	Fokker D.VII	Moulin -sous-Touvent.	
15.06.18:	*Cpl* Jean-Marie Martre KIA by *Leutnant* Carl Menckhoff, *Jasta* 72s.					
01.07.18	*Adj* Marinovitch	20	11	Rumpler	Monnes	
15.07.18	*Sgt* Crehore	21	2	2-seater	Vouziéres	
15.07.18	*Sgt* Crehore	–	Prob.	Balloon		
15.07.18	*Adj* Marinovitch	22	12	EA		Moronvillers
	Sgt Martin		2			
	Sgt Grimouille		1			
15.07.18	*Lt* Carbonnel	23	1	Rumpler C	Moronvillers	
	Adj Marinovitch		13			
	MdL Ducornet, Spa.93		2			
22.07.18	*Adj* Marinovitch	24	14	Rumpler C	Epernay	
	MdL Bessières		2			
02.08.18	*Lt* Bozon-Verduraz	–	Prob.	EA		
09.08.18:	*Sgt* André Martin KIA Ligniéres by *Ltn* Franz Büchner, *Jasta* 13.					
17.08.18:	*Adj* Marinovitch	25	15	Rumpler	Roye	
17.08.18:	*Adj* Marinovitch	26	16	Fokker D.VII	Roye	
17.08.18:	*Lt* Bozon-Verduraz	27	9	2-seater	NE Roye	
19.08.18:	*MdL* Maurice Caulier POW S Roye by *Ltn* Alfred Fleischer, *Jasta* 17.					
21.08.18	*S/Lt* Martenot	28	6	Fokker D.VII	W Soissons	*Vfw* Anton Bernhörster, *Jasta* 61, KIA
	Adj Ondet		1			
13.09.18	*Lt* Laganne	29	2	Fokker D.VII	Vieville-en-Haye	Possibly *LtdR* Paul Wolff, *Jasta* 13, WIA, POW
	Lt Carbonnel		2			
	S/Lt Martenot		7			
15.09.18	*MdL* Sainz	30	1	Balloon	Goin	
03.10.18	*Lt* Bozon-Verduraz	31	10	EA		
03.10.18	*S/Lt* Martenot	–	Prob.	Fokker D.VII	St. Martin l'Heureux	Possibly *Ltn* Fritz Höhn, *Jasta* 41, KIA
05.10.18	*Adj* Marinovitch	32	17	Fokker D.VII	E Challerange.	
09.10.18	*Lt* Bozon-Verduraz	33	11	2-seater	Bantheville	
	Lt Laganne		4			
09.10.18	*Sgt* Grimouille	34	2	Rumpler C	Semide	
	Lt Sardier, Spa.48		14			
	Sgt Denomaison, Spa.48		1			
09.10.18	*Adj* Marinovitch	35	18	Rumpler C	Semide	
09.10.18	*Brig* Froussard	36	1	2-seater	Machault	
12.10.18	*S/Lt* Martenot	–	Prob.	2-seater	E Tagnon	Possibly *Ltn* Fritz Guillaume & *Ltn* Bernard Schönleber, KIA, Annelles
14.10.18	*Adj* Ondet	37	2	Rumpler C	S Ecorday	
	Sgt Grimouille		3			
	MdL Sainz		2			
18.10.18	*Adj* Marinovitch	38	19	2-seater		
	Adj Ondet		3			
27.10.18	*S/Lt* Marinovitch	39	20	Fokker D.VII	Le Thour	*Vfw* Karl Schlegel, *Jasta* 45, KIA
28.10.18	*Lt* Laganne	–	Prob.	2-seater	St. Clement	Damaged
	S/Lt Martenot		Prob.			
28.10.18	*MdL* Sainz	40	3	Balloon	Hauteville	
29.10.18:	*Sgt* Henri Dussourd killed in crash.					
01.11.18	*MdL* Sainz	41	4	Balloon	Macheromesnil.	
03.11.18	*S/Lt* Marinovitch	–	Prob.	2 2-seaters	Le Chesne	
	MdL Froussard		Prob.			
03.11.18	*S/Lt* Marinovitch	42	21	2-seater	Le Chesne	
	Sgt Grimouille		4			
04.11.18	*Adj* Emile Picard WIA					

References

Frank W. Bailey and A.E. Clausen, Jr., "L'Escadrille de la 'Mort qui Fauche,' The Grim Reapers of Spa.94," *Cross & Cockade Great Britain Journal*, Vol.9, No.2, pp.112–120.

Frank W. Bailey and Christophe Cony, *The French Air Service War Chronology 1914–1918*, Grub Street, London, 2001.

Escadrille Spa.94 – journal des marches et operations, 1er Juin 1917-5 Novembre 1918.

Norman Franks, Frank Bailey and Rick Duiven, *Casualties of the German Air Service*, 1914–1920. Grub Street, London, 1999.

Norman L.R. Franks and Frank W. Bailey, *Over The Front: A Complete Record of the Fighter Aces and Units of the United States and French Air Services*, 1914–1018. Grub Street, London, 1992, pp. 101,189–190.

Dennis Gordon, *The Lafayette Flying Corps*, Schiffer Publishing Ltd., Atglen, PA, 2000, pp. 119, 373, 411, 477, 478, 480.

Email correspondence with June Gulick and Jack Gulick, 2015.

Charles Nordhoff and James Norman Hall, *History of the Lafayette Flying Corps*, Houghton Mifflin & Co., Boston, MA,1920, Vol.1, pp. 192, 200–202, 505–506, 509.

Daniel Porret, *Les "As" français de la Grande Guerre*, Service Historique de l'Armée de l'Air, Paris, 1983, Vol.I, pp.77–79, 321–322.

Greg VanWyngarden, *Jagdgeschwader Nr II: Geschewader 'Berthold,'* Osprey Publishing, Botley, Oxford, 2005.

Houston Woodward, *A Year for France: War Letters of Houston Woodward,* Yale Publishing Assn., Inc., New Haven, CT, 1919.

Left: The innovative Morane-Saulnier AI parasol monoplane served briefly at the front before wing failures ended its promising career and it was replaced by the more robust Spad fighters.

Right: The Morane-Saulnier AL was a development of the AI with a monocoque fuselage, 170hp Rhône 9R engine, enlarged fin, horn balanced elevators and twin Vickers machine guns. During its third test flight on May 16, 1918, it crashed, killing its pilot, five-victory ace Eugène Gilbert.

End Notes

The majority of quotations in this book are taken from a series of interviews the author had with André Martenot de Cordoux, as well as a running email correspondence in early 2015 with June Gulick, daughter of Austen B. Crehore, and his grandson, Jack Gulick. They are quoted as they appear in the text. Those taken from other sources are listed below:

Formation and First Blood

1. Jacques Mortane, "Les Héros Disparus: Henri Variot," *La Guerre Aérienne*, No.73, 4 April 1918, p.339.
2. *Ibid.*

Life at the Aerodrome

New Year Changes

3. Houston Woodward, *A Year for France: War Letters of Houston Woodward,* Yale Publishing Assn., Inc., New Haven, CT, 1919, pp.147–151.
4. *Ibid.*
5. Charles Nordhoff and James Norman Hall, *History of the Lafayette Flying Corps*, Houghton Mifflin & Co., Boston, MA,1920, Vol.1, p.202.
6. Nordhoff and Hall, p.200.
7. Woodward, pp.155–156.
8. Woodward, p.156.
9. Woodward, p.156.
10. Woodward, pp.157–158.
11. Acceuil Les "As" oubliés de 14–18, http.asoublies1418.fr/_zdocs/77/escadrilles/094.
12. Acceuil Les "As" oubliés de 14–18, http.asoublies1418.fr/_zdocs/77/escadrilles/094.
13. Adj. Walter John Shaffer, Letter to Mother, Plessis-Belleville, February 10, 1918, Unpublished Memoir, pp.234–235.
14. Adj. Walter John Shaffer, Unpublished Memoir, Letter to Mother, Secteur Postale 127, May 21, 1918, Unpublished Memoir, p.324.

Death Duel Over the Champagne

15. Acceuil Les "As" oubliés de 14–18, http.asoublies1418.fr/_zdocs/77/escadrilles/094.
16. Acceuil Les "As" oubliés de 14–18, http.asoublies1418.fr/_zdocs/77/escadrilles/094.
17. Woodward, pp.185–186.
18. Woodward, pp. 187–191.
19. Jon Guttman, "'Monsieur Moniteur," Cigogne: Interview with Louis Risacher,' *Cross & Cockade International Journal,* Vol. 20 No.2, Summer 1989.
20. Peter Kilduff, "'He Who Dares Wins,' A History of Royal Saxon Jagdstaffel 24," *Over the Front*, Vol.28, No.2, Summer 2013, p.148.
21. Peter Kilduff, *Hermann Göring Fighter Ace*, Grub Street, London 2010, p.137.
22. www.lesmortsdanslescamps.com/j-o.htm
23. Nordhoff and Hall, p.200.

Going Onto the Offensive

24. Nordhoff and Hall, p.200.
25. Norman L.R. Franks and Frank W. Bailey, *Over The Front: A Complete Record of the Fighter Aces and Units of the United States and French Air Services,* 1914–1918. Grub Street, London, 1992, p.190.
26. *Ibid.*
27. Dennis Gordon, *The Lafayette Flying Corps*, Schiffer Publishing Ltd., Atglen, PA, 2000, p.480.
28. Norhoff and Hall, p.509.
29. Nordhoff and Hall, p.199.
30. http://fandavioin.free.fr/Andr%20Henri%Martenot%20de%20Cordou.
31. htm Acceuil Les "As" oubliés de 14–18, http.asoublies1418.fr/_zdocs/77/escadrilles/094.

Month of Victory

32. Acceuil Les "As" oubliés de 14–18, http.asoublies1418.fr/_zdocs/77/escadrilles/094.
33. Acceuil Les "As" oubliés de 14–18, http.asoublies1418.fr/_zdocs/77/escadrilles/094.

Après-Guerre

34. A Geneology of Williams Families, via June C. Gulick.

Right: A career soldier since 1909 and an airman since 1916, *Lieutenant* Emile Deviterne assumed command of Spa.94 from an ailing Benjamin Bozon-Verduraz from November 18, 1918 through May 2, 1919. (Deviterne family via Albin Denis photo collection)

Made in the USA
Coppell, TX
12 November 2026

11803123R00057